MW00533386

The Expanding Universe

A Primer on Relativistic Cosmology

Cosmology – the science of the Universe at large – has experienced a renaissance in the decades bracketing the turn of the twenty-first century. Exploring our emerging understanding of cosmology, this text takes two complementary points of view: the physical principles underlying theories of cosmology, and the observable consequences of models of universal expansion.

- Includes a structured discussion of General Relativity, firmly based on conceptual foundations, with mathematics limited to the minimum necessary, enabling students to grasp the underlying physical principles.
- Relates modern observations to theories of cosmology, deriving and explaining the relationship between basic physical quantities and observations, to show how modern observational astronomy supports and informs cosmological theory.
- Discusses non-intuitive concepts based on the foundations of General Relativity and cosmology, supporting readers as they confront apparent paradoxes in modern cosmology.
- Carefully explains limitations on our current understanding of the Universe's structure and evolution, arising from incomplete physical theory and imperfect observations; with caveats as to possible future developments.
- Worked solutions to end-of-chapter problems are available online for instructors, via www.cambridge.org/heacox.

William D. Heacox is Emeritus Professor of Astronomy at the University of Hawaii at Hilo where he founded the undergraduate astronomy degree program. He has also had professional appointments at NASA, the University of Arizona, and Carter Observatory, and is an active member of the American Astronomical Society, International Astronomical Union, American Geophysical Union, and American Mathematical Society.

The Expanding Universe

A Primer on Relativistic Cosmology

William D. Heacox

University of Hawaii

CAMBRIDGE
UNIVERSITY PRESS

University Printing House, Cambridge CB2 8BS, United Kingdom

Cambridge University Press is part of the University of Cambridge.

It furthers the University's mission by disseminating knowledge in the pursuit of education, learning and research at the highest international levels of excellence.

www.cambridge.org
Information on this title: www.cambridge.org/9781107117525

First published 2015

Printed in the United Kingdom by TJ International Ltd. Padstow Cornwall

A catalogue record for this publication is available from the British Library

Library of Congress Cataloging in Publication data
Heacox, William D.
The expanding universe : a primer on relativistic cosmology / William D. Heacox, University of Hawaii.
pages cm
Includes bibliographical references and index.
ISBN 978-1-107-11752-5 (Hardback : alk. paper)
1. Expanding universe. 2. Cosmology. 3. Astronomy. I. Title.
QB991.E94H43 2015
523.1′8–dc23 2015024942

ISBN 978-1-107-11752-5 Hardback

Additional resources for this publication at www.cambridge.org/heacox

Dedicated to my mentors in astronomy:
the late Phyllis Hutchings (Whitman College);
the late Tom Lutz and Sid Hacker (Washington State University);
Ann Boesgaard and Sidney Wolff (University of Hawaii).
Mahalo nui loa.

Contents

Preface

The pace of change in cosmology has accelerated remarkably in the years bracketing the turn of the twenty-first century, so that many of the classical texts are becoming dated. The purpose of this text is to provide a coherent description of current theory underlying modern cosmology, at a level appropriate for advanced undergraduate students. To do so, the book is loosely organized around two pedagogical principles.

First, while the development of physical cosmology is heavily mathematical, the book emphasizes physical concepts over mathematical results wherever possible. The mathematics of General Relativity and of relativistic cosmology are beautiful, elegant, and seductive. It is a real temptation to develop theoretical cosmology as a purely mathematical structure, much as can be done with classical thermodynamics. But to do so is to lose sight of the deeper meaning of cosmology and to leave the student unprepared for the changes in the field that are almost certainly coming. In Einstein's inimitable phrasing,[1] *"Mathematics is all very well, but Nature leads us by the nose."* The book endeavors to lead the student gently, if not always easily, toward a useful understanding of the physical underpinnings of modern cosmology.

Cosmology is an inherently uncertain science, because of both the remoteness (spatial and temporal) of its subjects and the incompleteness of its observational foundations. It is thus not surprising that recent technological advances in observational astronomy have produced something of a revolution in cosmological theory, from inflation to dark energy to new theories of galaxy origins. But interpretations of cosmological observations are typically based on conceptual models and (in some cases) underlying physics of uncertain validity, so wherever possible the book derives and interprets its results in a manner conducive to

[1] "Die Mathematik ist shōn und gut, aber die Natur führt uns an der Nase herum"; in a postcard to Hermann Weyl dated 26 May 1923, and reproduced on page 83 of Nussbaumer and Bieri (2009).

re-interpretation when new observations and/or physics so permit. The book is also at some pains to point out the uncertainties of cosmology theory arising from incomplete observational evidence and adoption of specific physical models. Modern cosmology is truly an intellectual wonder, but is likely to experience considerable revision as new observations and physics come to bear upon it. This book will, hopefully, prepare students for such changes.

The choice of subjects to include in a text of reasonable length is largely a personal one, guided by perceived needs and the ready availability of other books and reference material. This book's main emphasis is on development of cosmological models describing the Universe's expansion, in a form that can be applied to current observations and that should be useful when and if new observations compel changes to current models. Missing from the book are extensive discussions of particles and quantum fields in the very early Universe; of the finer details of observational cosmology; and of related issues such as black holes and gravitational radiation. There is no shortage of other books covering these areas.

Conspicuously included here is the General Theory of Relativity (GR) up to the development of the Einstein Field Equations. This requires a substantial detour from cosmology *per se*, but in the author's opinion the effort required is justified in terms of deeper understandings of the physics of cosmology. General Relativity is arguably the most elegant and beautiful physical theory accessible to undergraduate students, and one to which all prospective physicists and astronomers should be introduced. The text endeavors to develop GR in a manner that avoids needless abstraction and that emphasizes the physical principles involved. Some of the more abstract and mathematical aspects of GR are relegated to Appendix A, for the edification of mathematically ambitious students.

The book's intended audience is advanced undergraduate physics and astronomy students, but much of the material can be understood by those with only modest preparation in these fields. The necessary mathematics background is limited to elementary calculus and differential equations, with a smattering of linear algebra in the form of elementary matrix manipulations. For most newcomers to the subject of GR, the most intimidating chapter will probably be Chapter 4, which describes tensor analysis at an elementary level. Tensor analysis has gotten a bad reputation because of its sophisticated applications, but at the level needed here the subject is surprisingly accessible. Very little actual tensor manipulation is required to understand GR at this level; mostly, students need only to become familiar with tensor notation. Working the problems and examples should help.

The text assumes that readers are familiar with classical and modern physics at the level usually presented in lower-division physics courses for physics and astronomy students in American universities. It would probably also help to have some knowledge of modern astronomy at an elementary level, but little of the subject will be lost to those coming to astronomy for the first time.

The values of constants and of conversion factors, and lists of symbols used in the text, appear as appendices. The natural cosmological units are giga-parsec for distance (1 Gpc = 10^9 parsecs $\approx 3 \times 10^{25}$ meters), giga-year for time (1 Gyr = 10^9 years $\approx 3 \times 10^{16}$ seconds), and solar mass (1 $M_\odot \approx 2 \times 10^{30}$ kg) for mass; these units are used throughout the text. Unless otherwise noted, section number references in the text include the chapter number: e.g., Section 9.3.1 is Subsection 1 of Section 3 of Chapter 9.

Problems for students to solve are included at the ends of most of the chapters: these range from fairly simple to rather involved, and are chosen for their pedagogical value. Students, and those reading the book for independent study, are encouraged to do them all. The book is at some pains to include references in the current literature to more extensive discussions of difficult or complex subjects, as guides for further study. Brief historical notes appear where appropriate, but the text makes no pretence of being a guide to the history of cosmology. For extensive discussions of that fascinating subject see, e.g., Nussbaumer and Bieri (2009), Ostriker and Mitton (2013), and Ferreira (2014).

Introducing the Universe

Cosmology is the study of the Universe on the largest scales. As such it deals with structures and dynamics that are quite different from those of terrestrial physics or, indeed, of most other branches of astronomy. It is a subject in which the finite speed of light plays a major role, conflating distance with time: in observing distant galaxies we see them not as they are now, but as they were when the light left them; and that is typically long enough ago to encompass significant evolution in the Universe's contents and dynamics. On cosmological scales the dynamics are dominated by gravitation and, possibly, dark energy. The current theory of gravitation is Einstein's General Theory of Relativity (GR), a non-linear and hugely complex theory entailing subtle and largely unfamiliar mathematical and physical concepts; the nature of dark energy remains speculative as of this writing. It is thus a non-trivial matter to assemble a coherent physical model of the Universe at large, requiring careful definitions of such seemingly mundane things as *distance* and *time*. We begin the effort in this introduction with a brief description of the Universe and its contents, and an assessment of our ability to understand the Universe on large scales in both space and time.

Galaxies and friends

Figure 1 illustrates the Universe as we naively think of it: a vast, crowded collection of galaxies. But this is deceptive, for such a picture collapses three dimensions into two and amplifies brightnesses, and thus under-represents the distances between galaxies and overstates both the density of matter and the degree of illumination of the Universe at large. The Universe in actuality is quite thinly populated and only faintly illuminated.

The visible contents of the Universe at large are almost entirely in the form of galaxies, mostly as large, luminous galaxies such as ours. Large galaxies, such as our Milky Way Galaxy, contain billions of stars and much diffuse matter. A typical example is shown in Figure 2: the spiral disk is ~ 25 kilo-parsecs

Figure I Hubble Extreme Deep Field (XDF) image, covering $\sim$ 4 square minutes of arc and containing images of $\sim$ 5500 galaxies. As with most digital images, the dynamical range is much higher than can be reproduced in a printed image, so only a portion of the field's contents are evident here. (NASA/ESA/G.Illingworth *et al.*/HUDF Team)

(kpc) in diameter; stars within it are, on average, $\sim$ 1 parsec (pc) apart. On this scale, only the very brightest stars are individually visible; stars similar to, or less massive than, the Sun are too faint to resolve in all but the very nearest galaxies. Galaxies are (nearly) self-contained dynamical and chemical systems and are, in effect, the proper inhabitants of the Universe. Large galaxies such as ours are probably outnumbered by much smaller galaxies, but still account for most of the Universe's stellar content and mass. Relatively very large galaxies, such as the giant ellipticals found at the centers of many galaxy clusters, are so rare as to be relatively minor contributors to the overall galaxy population.

The total mass density of galaxies, averaged throughout the Universe, is the equivalent of less than 1 baryon (proton, neutron) per cubic meter. This is incredibly thin: 'empty' space in our Solar System, and in interstellar space in our part of the Galaxy, is denser by at least four orders of magnitude. By human standards the Universe is very nearly empty. One consequence is that luminous matter is thinly spread throughout the Universe, which is thus only dimly illuminated by the stars that account for nearly all the Universe's visible radiation. The average luminous flux in the Universe is the rough equivalent of that of a 100-watt light bulb at a distance of 10 km, effectively undetectable by the unaided human eye if at all diffuse. By human standards the Universe is, overall, very dark.

Figure 2 Spiral galaxy M101 (NASA/HST).

You can verify this simply by looking up on a dark night: visible stars are so numerous and bright that, at a good site in the right season, the stars in the Milky Way cast a visible shadow. But galaxies external to ours are so faint that, unless you know just where to look and have a very dark sky, you will not see any. The only external galaxy (not a satellite of our Galaxy) that is at all likely to be seen with unaided human vision is M31, the Andromeda Galaxy; barely visible in the northern winter skies as a faint, fuzzy patch. Get out between galaxies in an average location in the Universe and your sky would contain few, if any, visible objects. The bright glory of our night skies is an artifact arising from our quite atypical location in the Universe: relatively *very* dense and bright.

Most of the Universe's galaxies are independent or nearly so, but about 10% are organized into a hierarchy of gravitationally mediated structures ranging in size from small ($\lesssim$ 10 galaxies) groups such as the Local Group in which our Galaxy is located; to larger clusters containing hundreds or thousands of galaxies, as illustrated in Figure 3. On even larger scales are clusters of clusters – 'superclusters' – containing up to 10^5 galaxies and spanning 100–150 million parsecs (Mpc). Beyond this there are no larger structures (that we can perceive): the Universe appears to be very smooth on scales greater than $\sim$ 150 Mpc, with typical density variations limited by $|\delta\rho/\rho| \lesssim 10^{-5}$.

In our part of the Universe – the Local Group – large galaxies are typically separated by $\sim$ 1 Mpc, or several tens of their diameters; in larger groupings the galaxy densities can be much greater and galaxy–galaxy interactions are probably common. The total number of large galaxies in the visible Universe is on the order of $\sim 10^{12}$, each of them containing $\gtrsim 10^{10}$ stars. A very large Universe,

Figure 3 Abell 2744, a large cluster of galaxies (NASA/HST.)

indeed; but probably only a tiny portion of the total number, limited as it is by the finite speed of light: since the Universe is of a finite age, only those galaxies close enough for light to have reached us are visible.

Universal expansion

Modern cosmology can fairly be said to have started with the discovery that galaxies appear to be receding from us at speeds proportional to their distances, a large-scale feature that is commonly interpreted as expansion of the entire Universe. The conceptual picture here is one of a uniformly expanding substrate that carries galaxies along with it, so that all galaxies are receding from all other galaxies at speeds proportional to distance: think of ink spots on an expanding balloon, or raisins in expanding bread dough.

The dynamics of such an expansion are best studied as those of a continuum characterized by a universal and non-dimensional expansion function $a(t)$, in terms of which the distance d between any two galaxies evolves as

$$d(t) = a(t)\, d_0 \,, \tag{I.1}$$

where $d_0 = d(t_0)$ is the current distance and a is normalized to $a(t_0) = 1$ (t_0 is the current time). The relative radial velocity between the two galaxies is then $V = \dot{d} = \dot{a}\, d_0 = (\dot{a}/a)\, d$, or velocity proportional to distance. The relative expansion rate $H \equiv \dot{a}/a$ is the **Hubble Parameter**, named for the eponymous astronomer

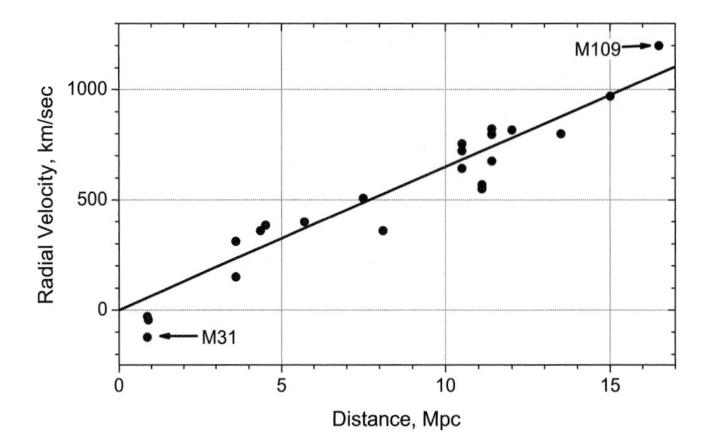

Figure 4 Velocity–distance relation for nearby bright galaxies. The straight line is a least-squares fit with slope 65 km/sec/Mpc (the Hubble Constant for nearby galaxies; see text). Two Messier catalog galaxies are identified with their Messier numbers.

who first proposed the relation

$$V = \frac{\dot{a}}{a} d = Hd \,, \tag{I.2}$$

for galaxies at cosmological distances. The units of H are properly those of inverse time but, following Equation (I.2), are usually given in the mixed units 'km/sec/Mpc'. Since both d and V are observable properties of galaxies, this **Hubble Relation** may be used to estimate the Hubble Parameter and thus the expansion rate of the Universe.

The reality of this Hubble relation is demonstrated in Figure 4, which is similar to that originally published by Hubble in the 1920s (but with more modern data). There is a fair amount of scatter here, due mostly to local effects such as mutual gravitational attractions amongst galaxies and at least partly to uncertain distance estimates; but the obvious trend is compelling evidence of an overall expansion. From such observational evidence for large numbers of galaxies, we estimate the current value of the Hubble Constant $H_0 \equiv H(t_0)$ to be $\approx 72 \pm 5$ km/sec/Mpc, or ≈ 0.074 Gyr^{-1}.

Universal evolution

The Universe shows evident signs of evolution of its content (principally in its galaxy population) and contains radiation interpreted as the remnant of an early, hot phase of the entire Universe. The current theory encompassing these matters is the 'Big Bang' theory, in which the Universe starts in a very hot, dense state from which it expands and cools. With addenda from particle and quantum physics, the Big Bang theory constitutes the current cosmological paradigm.

If the Universe were expanding at a constant rate $\dot{a}$ beginning with $a(0) = 0$, the expansion function would evolve as $a(t) = \dot{a}t$ and thus $H = \dot{a}/a = 1/t$; and the current age would be the **Hubble Time**, $t_H = 1/H_0$, a characteristic time scale in an expanding universe. With our observational estimate of $H_0 = 0.074\ \text{Gyr}^{-1}$ the Universe's current age would then be $t_0 = t_H \approx 13.5$ Gyr. This is gratifyingly similar to the estimated age of the oldest stars in our Galaxy, and suggests that our Galaxy formed early in the Universe's expansion. But gravitation and dark energy will modify the expansion rate, so such age estimates can only be approximations.

After its initial startup phases, and prior to the age of $\sim$ 400 million years, the Universe was filled with a diffuse, nearly uniform hydrogen/helium gas lit only by its own thermal radiation. As stars and galaxies formed from this gas, and the Universe expanded, it became less homogeneous but diffuse overall, and illuminated by starlight: at the time of maximum star formation rate ($t \sim 3$ Gyr) it was more than an order of magnitude denser than it currently is, and brighter by several orders of magnitude. Since that distant time the Universe has become thinner and darker, and will probably continue to do so into the foreseeable future.

Universal dynamics

The Universe's evolution is a dynamical consequence of its initial expansion – inherited from the Big Bang – and of its matter and energy contents. Since Special Relativity has it that energy and matter are equivalent, both are sources of gravitation and contribute to the Universe's dynamical history. That history is dominated by exotic forms of matter and energy – generically labelled 'dark' – of currently unknown origin and nature, that constitute $\sim$ 95% of the Universe's gravitating contents. What we observe as the visible Universe (galaxies, etc.) constitutes a sort of thin veneer over the 'real' – but unobserved, and mysterious – Universe of dark matter and energy. Observations of the effects of dark energy and matter on the visible Universe, and reasonable extrapolations from known physics, hold out hope that the dynamics of these forms of matter/energy can be understood and modelled, even if the dark stuff itself remains of a mysterious nature. Models of the Universe's expansion, such as are developed in this text, are based on such expectations which may be proven wrong in the fullness of time, and require correction. The search for the nature of these dark components is a major effort of modern cosmological research.

Universal models

Cosmologists attempt to explain large-scale features of the Universe, such as the Hubble Relation or its hierarchical structure, in terms of physical models that are based largely on explicit forms for the expansion function $a(t)$ as defined in Equation (I.1). On cosmological scales the dominant forces are those of gravitation, for which the currently accepted theory is Einstein's General Theory of Relativity (GR); and dark energy, for which there is currently no good theory but whose dynamical effects appear to be understood from observations of the Universe's expansion. Applied to a Universe of given distributions of mass and energy, GR should produce a model for $a\,(t)$ whose observational consequences would be testable, much as Newtonian gravitation predicts the paths of objects subject to gravitational fields produced by distributions of masses.

The cosmologists' game then becomes one of estimating the proper form of the expansion function by working the problem from both ends. On the one hand we estimate the mass/energy densities as closely as possible from direct observations of, e.g., galaxy masses and spatial densities, and electromagnetic radiation fields; and compute $a\,(t)$ therefrom, using GR to relate expansion to gravitation. On the other, the observational consequences of the models, such as the Hubble Relation between distance and recession velocity, are tested against actual observations in order to assess their validity. Neither of these approaches gives unambiguous results since they both rely on difficult observations of uncertain accuracy, but iteration on the pair typically converges to a self-consistent model for the expansion function that is of useful validity, to the extent that the underlying observations are correct.

The remainder of this text sets up the physics needed to model the expanding Universe (Part II) and then uses the results to construct physically meaningful models of the Universe's structure and dynamics (Parts III–V). Of necessity, the discussion is rather abstract and heavily mathematical in places, so we devote Part I to easing the student into the subject with an introduction to the underlying concepts.

Part I

Conceptual foundations

1

Newtonian cosmology

While General Relativity (GR) is the only theory of gravitation that can describe the Universe on large scales, Newtonian gravitation provides an illuminating approximation to relativistic cosmology. It is thus worthwhile to briefly consider classical, non-relativistic Newtonian gravitation and its application to cosmology, before diving into the complexity and abstraction of the general relativistic version.

1.1 Newtonian gravitation

Newton's theory of gravitation has been spectacularly successful in application to all sorts of problems over the last 300+ years. But it has conceptual shortcomings at its roots that render it inapplicable to all but the simplest approximations to reality when applied to strong gravitational fields, or on the scales encountered in cosmology.

First, the theory includes no mechanism for the transmittal of gravitational force from one mass to another. The theory was criticized for this 'action at a distance' requirement during Newton's lifetime, but he brushed off the criticisms with the observation that the theory worked even if we did not know exactly how. The large distances encountered in cosmological applications, however, require a theoretical underpinning that accommodates the transmittal of gravitational forces over billions of light years; and the apparent instantaneous nature of Newtonian gravitation over such distances seems suspect at best.

Second, Newtonian gravitation offers no explanation for the equality of gravitational and inertial mass. The gravitational force exerted by a mass M on a massive test object located a distance d away is

$$\vec{F} = -G\frac{Mm_{(G)}}{d^3}\vec{d} \, ,$$

where by $m_{(G)}$ is meant the gravitational mass of the test object, that appearing in Newton's law of gravitation (above). The dynamical response of the test object is the acceleration

$$\vec{a}_m = \frac{\vec{F}}{m_{(I)}} = -G\frac{M}{d^3}\frac{m_{(G)}}{m_{(I)}}\vec{d}\,,$$

where $m_{(I)}$ is the inertial mass of the object, that appearing in Newton's second law of mechanics, $\vec{F} = m\vec{a}$. It is a well-observed fact that $m_{(G)} = m_{(I)}$ to a high degree of precision *for all massive objects,* so

$$\vec{a}_m = -G\frac{M}{d^3}\vec{d}\,,$$

independent of the object's mass: a gravitational field accelerates all massive objects at the same rate, irrespective of their mass (or anything else).[1] Gravitation is the only fundamental force for which this is true, and Newton's theory offers no suggestion as to why this is so.

Third, Newtonian dynamics (including gravitation) is based on concepts of absolute space and time. Newton's second law ($F = m\vec{a}$), for instance, only works in inertial reference frames, those experiencing no acceleration. But acceleration relative to *what*? If you're the only thing in the Universe, how do you know if you're accelerating or not? The nineteenth century Austrian physicist Ernst Mach had an interesting answer to such questions, since embodied in Mach's Principle:[2] that the origin of inertia lay in the combined gravitation of all the Universe's contents, and that space of itself had no existence as a thing. This principle was an important one in Einstein's thinking leading to GR, but has since fallen out of favor with modern physicists who include fields as properties of space itself.

But for Newton the only answer was that there was an absolute space relative to which all accelerations could be measured. Similarly (if not quite so obviously), there must be an absolute time that applies to all of space. Einstein's Theory of Special Relativity (SR) showed that both of these absolute concepts were erroneous, thus largely undermining the fundamentals of Newtonian gravitation.

Fourth, Newton's theory of gravitation is manifestly incorrect in applications to strong gravitational fields. At the time Einstein took on the task of developing its successor the most worrisome and well-established discrepancy in Newtonian

[1] The subject of Galileo's (possibly apocryphal) experiment of dropping objects of different weights from the Leaning Tower of Pisa, to see if they hit the piazza below at the same time. A more compelling version of the experiment, without the complication of air resistance, was performed by Apollo astronauts on the Moon using a rock hammer and a feather as test objects.

[2] See, e.g., Sciama (1969), Graves (1971), Rindler (1977), Peebles (1993), and Ghosh (2000) for analysis of Mach's Principle and discussions of its relation to GR and its present standing in modern physics and cosmology.

gravitation had to do with the orbit of Mercury. That elliptical and inclined orbit, embedded deep within the Sun's gravitational field, precesses at a rate of $\Delta\theta = 5599''.74 \pm 0''.40$ per century, due mostly to the gravitational perturbations of other planets. But all attempts to model the precession from Newtonian gravitation fell short by about $43''$ per century, a figure two orders of magnitude larger than the estimated observational uncertainty. By the beginning of the twentieth century this apparent failure of Newtonian gravitation had become sufficiently worrisome as to prompt several attempts to modify the theory itself. Einstein's demonstration that his General theory of Relativity successfully predicted this anomalous precession was an important factor in its reception by the scientific community. Observations during the intervening ~ 100 years have since revealed several other areas in which GR gives a correct answer where Newtonian gravitation does not (think: black holes!).

1.2 Universal expansion

Yet another problem with Newtonian gravitation arises in application to dynamical models of the Universe as a whole; i.e., cosmology. A gravitationally mediated expansion characterized by densities of gravitating matter and energy can be described by two sets of equations: one or more field equations relating mass/energy densities to gravitational potentials, and the resulting equation of motion for a test particle in that potential. In Newtonian mechanics these are, respectively,

Newtonian Gravitation

Field (Poisson's) Equation: $$\nabla^2\Phi = 4\pi G\rho \ , \tag{1.1}$$

Equation of motion: $$\vec{\mathbf{a}} = -\vec{\nabla}\Phi \ , \tag{1.2}$$

where ρ is the mass density, Φ is the Newtonian gravitational potential, and $\vec{\mathbf{a}}$ is the acceleration of a test particle in the gravitational potential. But Newtonian gravitational potentials cannot be unambiguously defined in an infinite, homogeneous medium where, by symmetry, the potential must be the same everywhere. Since Φ has no gradient under such conditions, there can be no gravitational dynamics. And since current observations strongly support the world-view of an effectively infinite and homogeneous[3] Universe – as did conventional thinking

[3] Homogeneous on sufficiently large scales.

prior to the development of GR and modern astronomical observations – this constitutes a serious obstacle to applying Newtonian gravitation to cosmology.

The matter can be finessed by suitable adjustments to the Newtonian theory, usually in terms of changes to Poisson's Equation.[4] But for purposes of illustration and comparison with the relativistic model to be derived in Chapter 8, it suffices to confine the analysis to finite universes with centers and to employ a classical energy analysis. Thus: the total mechanical energy E of a mass m a distance d from a fixed, central mass M is

$$E = U + K = -G\frac{Mm}{d} + \frac{1}{2}m\dot{d}^2 ,$$
$$\Rightarrow \dot{d}^2 = 2G\frac{M}{d} + 2(E/m) .$$

We can cast this into a cosmological context by (1) re-introducing the universal expansion function $a(t)$ (Equation (I.1)) so that $d(t) = d_0\, a(t)$; and (2) replacing the central mass M with uniformly distributed mass density throughout the spherical volume encompassed by the two masses, so that $M = (4/3)\,\pi\,\rho d^3$. Using these relations and the mass conservation condition $\rho d^3 = \rho_0 d_0^3$ to eliminate M and d from the above energy equation yields a differential equation for the expansion function:

$$\frac{\dot{a}^2}{a^2} = \frac{8\pi G}{3}\frac{\rho_0}{a^3} + 2\frac{E/m}{a^2 d_0^2} . \tag{1.3}$$

Note that the left-hand side of this equation is the square of the Hubble parameter, $H \equiv \dot{a}/a$, as defined in the Introduction.

This is the equation of motion of an object falling upward in a static gravitational field. The form of its solution depends critically on the value of E. If $E > 0$ the structure is unbound: $\dot{a}^2 > 0$ at all times and the expansion may continue forever (at an ever-decreasing rate). This is an **open** expansion corresponding to velocities exceeding that of escape. But if $E < 0$ the structure is gravitationally bound: $\dot{a} \to 0$ at sufficiently large a and the expansion stops and reverses itself. This is a **closed** expansion corresponding to velocities less than that of escape. The **critical** case separating these two corresponds to $E = 0$ and represents an object with exactly the escape velocity. Graphical examples of $a(t)$ for all three cases are shown in Figure 1.1.

The fate of this Newtonian Universe can be discerned by comparing its mass density to its expansion rate. From Equation (1.3) at the current time (when $a = 1$),

$$E = 0 \quad \Rightarrow \quad H_0^2 = \frac{8\pi G}{3}\rho_0 ,$$

[4] See Section 9.2 of Rindler (1977) for examples.

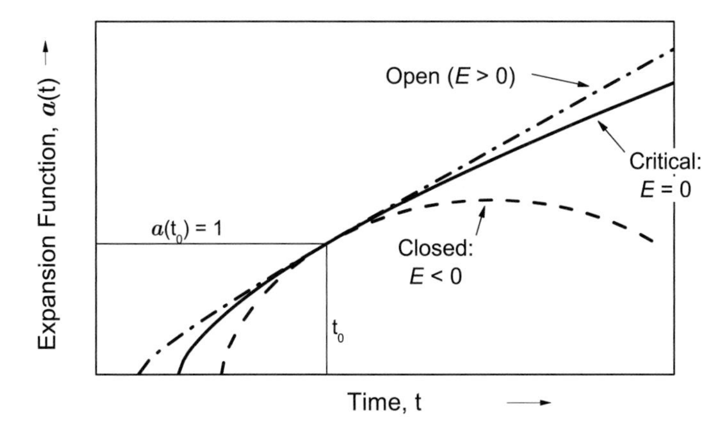

Figure 1.1 Sample expansion functions for Newtonian mechanics, all for the same value of H_0 (which is the slope of these functions at the current time, t_0).

where $H_0 = (\dot{a}/a)_{t_0} = \dot{a}(t_0)$ is the current value of the Hubble Parameter and $\rho_0 = \rho(t_0)$ is the current mass density. The critical mass density – that required for a zero-energy Universe currently expanding at the rate H_0 – is thus

$$\rho_{c,0} \equiv \frac{3H_0^2}{8\pi G} . \tag{1.4}$$

Inserting this into the expansion equation (1.3) for the current time and solving for the total energy:

$$2\frac{E/m}{d_0^2} = H_0^2\left(1 - \frac{\rho_0}{\rho_{c,0}}\right) . \tag{1.5}$$

The Newtonian Universe is open ($E > 0$) or closed ($E < 0$) depending upon whether its current mass density is less, or greater, respectively, than the critical density. Low density universes expand forever, high density ones eventually stop expanding and re-contract.

This is about as far as it is useful to carry the Newtonian analysis of cosmological expansion. While Equations (1.3) and (1.5) are gratifyingly similar to the fully relativistic expansion equations to be developed later in this book, and it is possible to join this Newtonian result to SR kinematics so as to produce a coherent picture of such things as cosmological redshifts; the logical sleights of hand employed in derivation of the Newtonian expansion equation largely invalidate its application to the real world. We need to invoke GR in order to realistically model the Universe on large scales, and that will require the considerable machinery developed in Parts II and III of this text.

Problems

1. Show that solutions to the expansion equation (1.3) in the critical case ($E = 0$) are of the form $a(t) \propto t^{2/3}$. Find an expression for the current time t_0 in such models in terms of the current mass density, ρ_0. Estimate the age of the Universe in this model in Gyr (10^9 years) if its mass density corresponds to 1 baryon (proton, neutron) per cubic meter, which is approximately what is currently observed.
2. Find an expression for the Hubble Parameter $H = \dot{a}/a$ as a function of time for the model of Problem 1. Find a numerical value for H for the current time and mass density of Problem 1, in units of both Gyr^{-1} and km/sec/Mpc.
3. Differentiate the Newtonian expansion equation (1.3) to derive an acceleration equation of the form $\ddot{a}/a$ as a function of the expansion function a, independent of the total energy. Show that this corresponds to a pure force of attraction.

2

General Relativity

Einstein's General Theory of Relativity (GR) was motivated principally by his desire to expand his very successful theory of Special Relativity (SR) to non-inertial reference frames. SR served to reconcile the invariance of the speed of light for all observers – as predicted by Maxwell's Equations of electromagnetism, and verified by the Michelson–Morley experiment – with Einstein's Principle of Special Relativity: that the laws of physics were the same in all non-accelerating reference frames. General Relativity, as Einstein envisioned it, would require the laws of physics to be identical in *all* reference frames, including accelerating ones. That this extension of the relativity principle leads to a theory of gravitation – which is what GR has become – was a consequence of the observed equality of gravitational and inertial mass: since *all* objects fell with the same acceleration in a given gravitational field, acceleration and gravitation are, in some sense, equivalent. Note that this singling out of gravitation distinguishes it from other fundamental forces, such as electromagnetism: acceleration and gravitation are connected in a unique manner.

But the details of that connection were totally non-obvious when Einstein set out to discover them; in particular, it did not seem possible at first to write laws of mechanics in a manner that is independent of the acceleration of the reference frame. In fact, Einstein never successfully united *all* forms of non-inertial motion into a single theory, but he did manage to do so with gravitation so that his General Relativity theory has effectively become one of gravitation, relegating Newton's theory of gravity to that of an approximation to the full relativistic theory. In particular, it is Einstein's theory of gravity that must be employed on the scales encountered in cosmology for a successful theory of the Universe's large-scale structure and evolution to be constructed.

The fundamental concepts underlying Einstein's theory of gravitation are these three: **General Covariance**, which expresses the relativity principle, that the laws of physics take the same form in all reference frames; **Equivalence**, which embodies the equality of gravitational and inertial mass; and **Space-Time**

Curvature, which provides the means by which gravitation controls dynamics. These are conceptually summarized in this chapter and are each the detailed subject of a separate chapter in Part II of this text.

2.1 Covariance

The Principle of General Covariance, as Einstein expressed it, is that the laws of physics are independent of our choices of reference frames or of coordinate systems, and that the equations of physics, properly constructed, should take the same form in all coordinate systems. Insistence on this property in development of SR proved to be crucial in extending SR to all of physics, including electromagnetism, in inertial reference frames. What worked so successfully in SR would apparently be a good choice for development of GR, at least in the basic stages.

The most obvious way to free physical laws from specific coordinate systems is to make all expressions of physical quantities overtly independent of coordinates. Thus, the Newtonian expression for gravitational potential Φ can be concisely written as Poisson's Equation: $\nabla^2\Phi = 4\pi G\rho$, which is true in all coordinate systems. But such simple representations are not readily extendable to mechanics in general, so instead we employ *co*variant forms which, while changing with coordinate systems, do so all in the same manner so that mathematical expressions of equalities of physical quantities remain unchanged even when coordinate systems change. Thus: if we have, say, a generally covariant vector equation of the form $A_i = B_i$ in one coordinate system and we change to another system, so that $A_i \to A'_i$ and $B_i \to B'_i$, it will nonetheless remain that $A'_i = B'_i$ as in the original coordinate system. We say that generally covariant vector components are *not in*variant – they *do* change with changes in coordinate systems – but they *are co*variant – they all change in the same manner so as to preserve their equality.

Equations written entirely in terms of generally covariant quantities remain true in all coordinate systems, including accelerating ones.

That general covariance is a special quality can be seen by considering the equations of motion of a force-free particle: $d^2x^i/dt^2 = 0$ for all coordinates x^i. This equation is manifestly untrue in, say, a rotating coordinate system defined by $\bar{x}^i = x^i\cos(\omega t)$, so that the simple equations of Newtonian mechanics are not generally covariant. To make them so we must write Newton's laws as, e.g., $\vec{\mathsf{a}} = 0$ in which $\vec{\mathsf{a}}$ assumes different (implied) forms in different coordinate systems: in accelerating systems this would include such complications as Coriolis and centripedal accelerations.

It turns out that all equations of physics may be written in a generally covariant form if one is willing to accept very complicated expressions, but as a practical matter it is better to adopt generally covariant mathematical forms that occur sufficiently frequently in a natural manner so as to be available for most applications; and are simple enough to be readily manipulated. The first such forms to be developed were based on the invariance of such physical/mathematical objects as normal vectors to a surface, or invariant measures of distance. These are primitive forms of **tensors**, and their use in such applications as GR is called *tensor analysis*. Other means of accomplishing the same goals as general covariance have since been developed, particularly in the area of modern differential geometry and associated subjects; but tensor analysis remains the least abstract and most commonly used form of covariance in physics. We will be discussing the needed parts of this subject in Chapter 4.

2.2 Equivalence

Einstein expanded the equality of gravitational and inertial mass into a Principle of Equivalence that has profound consequences for his theory of gravitation. The illuminating moment for Einstein came (he claims) in 1907 when, working at his study, he watched workmen on the roof of a nearby building and considered the hypothetical experience of one of them falling from the roof. He reasoned that during the course of the workman's fall to the ground he would feel no gravitational force: it would be as if he were floating in space, much like the videos of space shuttle and space station astronauts that we are all familiar with. The apparent weightlessness experienced by orbiting astronauts is a reflection of their freely falling in the Earth's gravitational field, together with everything in their effective reference frame (i.e. the space shuttle/station and all its contents). A reference frame falling freely in a gravitational field was effectively an inertial frame, within which the laws of special relativity were valid. Expanding on this idea in one of his famous *Gedankenexperiments* (thought experiments), Einstein surmised that a person locked in a cell with no windows could not distinguish between a gravitational field (the cell resting on the Earth) and uniform acceleration (the cell attached to a rocket accelerating at 1 g). Generalizing, he concluded[1] '*...the complete physical equivalence of a gravitational field and the corresponding acceleration of the reference frame.*' If this seems obvious now, observe that what Einstein did was to generalize a result that was manifestly true for mechanics alone, to all of physics (including electromagnetism, optics, thermodynamics, etc.). It was an intellectual leap of faith – not at all verified by observations other than of simple mechanical experiments – that at once converted his General Relativity into a theory of gravity.

[1] In Über das Relativitätsprinzip und die aus demselben gezogenen Folgerungen, *Jahrbuch der Radioaktivität und Elektronik*, **4**, 1908, p. 411. Translated by E. Osers and quoted by A. Fölsing in *Albert Einstein*, Penguin, 1998 (p. 303).

The special importance of this principle to GR is that it allows one to 'transform gravity away' by the simple expedient of changing coordinate systems to one freely falling in the gravitational field. There the appropriate (and generally covariant) relations can be derived as if in an inertial reference frame, whereupon one can transform back into the original coordinate system and obtain a correct formulation. That this strategy works is a testament to the power inherent in the use of generally covariant expressions for physical quantities. This approach to transforming gravitation into an accelerating – but internally inertial – reference frame is particularly valuable in setting up the expansion equations of cosmology, as we shall see in Chapter 8.

The Equivalence Principle has useful applications outside of GR *per se*: by ascribing effects observable in an accelerating reference frame to gravitation, one can deduce gravitational bending of light rays, gravitationally induced redshifts, and gravitational time dilation. These are discussed further in Chapter 5; Cheng (2005) gives a particularly illuminating and detailed discussion of such consequences of the Equivalence Principle.

2.3 Curvature

The central issue in Einstein's development of his theory of gravitation is that of the origin of equivalence: how does gravitation cause all objects to accelerate the same? The key idea came to Einstein as a consequence of a problem that had arisen in the interpretation of SR around 1910: what does a stationary observer measure for the circumference of a rotating wheel? Rotation is a form of acceleration so this is not a straightforward application of SR; but since each little segment of the rim of the wheel must contract along the direction of its motion when seen by an observer at rest, the overall circumference must shrink. On the other hand, the spokes of the wheel – being perpendicular to the motion in all cases – will not suffer length contraction as seen by a stationary observer. This suggests that $C < 2\pi R$ for a rotating wheel, which would seem to conflict with the Euclidean formula $C = 2\pi R$, a paradox that remained unresolved until Einstein connected it to the geometry of space itself: this same result ($C < 2\pi R$) applies to circles drawn on the curved surface of a sphere, as you can readily verify by drawing a circle on a globe. Einstein thus postulated that the acceleration represented by the rotation of the wheel *changed the geometry of space itself*, so that the Euclidean formula for the circumference was no longer correct.[2]

[2] Einstein's logical path to this conclusion is detailed in Chapter 6 of T. Levenson, *Einstein in Berlin* (Bantam, 2003).

In conjunction with the Equivalence Principle this curvature conjecture leads to the connections

$$\text{gravitation} \longleftrightarrow \text{acceleration} \longleftrightarrow \text{geometry of space-time},$$

from which he concluded that **gravitating bodies alter the geometry of space-time**. A very bold step, typical of Einstein at his best and, as with the Equivalence Principle, an extension beyond what was experimentally verifiable at the time. This is the remaining piece to moving beyond Newtonian gravitation: the mechanism by which objects charge space-time with gravitation is the geometry of space-time itself; gravitating bodies change the geometry, and *all* objects move in response to that change, irrespective of their mass. The GR equivalent of Newton's equation of universal gravitation will thus be an equation connecting sources of gravitation to the curvature of space-time.

As elegant as this conjecture is, its practical application is beset with serious mathematical and conceptual difficulties. Exactly what sort of curvature arises from a particular distribution of matter and energy, and precisely how does the resulting curvature affect the dynamics of objects in curved space-time? An important key to the solution to this problem was the recognition that the manifest properties of space and time restricted the possible forms of curvature to relatively simple geometries. In particular, any physically useful geometry must admit of a means of measuring distance, and must be smooth enough to permit differentiation along paths in the curved spaces. These attributes, plus the requirement that any surface be apparently flat when viewed very closely (e.g., the surface of a very large sphere), were those of **Riemannian Geometries** as first explicated by the nineteenth-century mathematician George Friedrich Bernhard Riemann, a student of Gauss.

In Riemannian spaces the distance along any curve may be measured by integrating the differential distance, or line element, along the curve. Thus, on the surface of a sphere of radius R the length of a path is the integral $\int ds$ along the path, where

$$ds^2 = R^2 d\theta^2 + R^2 \sin^2\theta \, d\phi^2 \tag{2.1}$$

is the incremental distance between two nearby points separated by infinitesimal coordinate differences $d\theta$ and $d\phi$ (geographically, θ is the equivalent of colatitude, ϕ is longitude). Similarly, on a plane surface and using polar coordinates,

$$ds^2 = dr^2 + r^2 d\theta^2 \; .$$

The differences between these two distance measures arise from the differing geometries of the underlying surfaces.

2.3.1 Metric tensor (I)

In general, one may write for the incremental distance in any Riemannian curved space

$$ds^2 = \sum_{i,j} g_{ij} dx^i dx^j \;, \tag{2.2}$$

where x^i is the ith coordinate (e.g., r, θ, ϕ) and the quantities g_{ij} express the curvature of the space in which the path length is being measured. These are components of the **metric tensor**, a matrix generically denoted by **g** with indexed components $g_{ij} \equiv (\mathbf{g})_{ij}$. The metric tensor **g** is a characteristic of the space and coordinate system being used and is fundamental in GR: gravitating matter and energy establish the form of this tensor through the curvature they induce, and that form in turn determines the paths of freely falling objects. As an example, the metric tensor matrix for the surface of a sphere in the coordinates (θ, ϕ) used in Equation (2.1) is

$$\mathbf{g} = \begin{pmatrix} R^2 & 0 \\ 0 & R^2 \sin^2\theta \end{pmatrix} ,$$

so that $x^1 \equiv \theta, x^2 \equiv \phi$, and $g_{11} = R^2, g_{22} = R^2 \left(\sin x^1\right)^2$. Note that the components g_{ij} of the metric tensor may be functions of the coordinates themselves. This is true even in familiar coordinate systems in flat, Euclidean space: polar coordinates in three dimensions (r, θ, ϕ), for instance, have metric

$$\mathbf{g} = \begin{pmatrix} 1 & 0 & 0 \\ 0 & r^2 & 0 \\ 0 & 0 & r^2\cos^2\theta \end{pmatrix} .$$

In general, the metric tensor characterizes the geometry underlying the coordinate system and is thus a fundamental quantity in GR.

2.4 Relativistic gravitational dynamics

The triumph of Einstein's GR theory lay in his establishment of a concise relation between the distribution of sources of gravitation, and the metric tensor that described the resulting space-time curvature. Schematically, these **Field Equations of Gravitation** take the form

$$\mathcal{G}(\mathbf{g}) \longleftarrow \mathcal{T}(\rho, \varepsilon, P \ldots) \;, \tag{2.3}$$

where $\mathcal{G}$ (g) is a set of differential equations in the metric tensor components, which will include time as one dimension; and $\mathcal{T}$ expresses the distribution of sources of gravitation. This will include not only matter but, by SR, all forms of energy including electromagnetic radiation, kinetic energy, gas pressure, etc. The Newtonian analog of the field equations is Poisson's Equation, $\nabla^2\Phi = 4\pi G\rho$, which is a set of second-order differential equations for the coordinate components of the gravitational potential Φ. By analogy, the field equations should include second derivatives of the metric tensor components, some of which serve as analogs of gravitational potentials. But the field equations are non-linear where Poisson's equation is not, the non-linearity arising in part from the inclusion of self-gravitation in the relativistic theory: mass and energy produce the equivalent of gravitational potential energies which, being a form of energy themselves, induce additional gravitation. This makes the field equations formally consistent with SR but also adds complexity and makes them much more difficult to solve.

Einstein's inference of the field equations was an inspired work in modern differential geometry and physical intuition. There are an unlimited number of mathematically plausible ways in which one can relate the metric tensor components to sources of gravitation. In choosing among them for his field equations Einstein was guided by analogy with Newtonian physics; by the requirement that his equations reduce to Newtonian gravitation in the appropriate weak-field limit; and by the partially philosophical principle ('Occam's Razor') that simple equations are more likely to be correct than are complicated ones. His field equations have withstood the tests of time but are still of uncertain validity at some level: they were not written in stone on Mt. Sinai, after all, and remain subject to possible future corrections.

The remaining piece of GR dynamics is that of the trajectory followed by an object freely falling in a gravitational field. In Newtonian mechanics a force-free object follows a straight line at constant velocity, which is the shortest distance between two (space-time) points. The analogy in curved space-time induced by gravitation is that of a **geodesic**, the shortest distance between two points. On the surface of a sphere, for instance, this would be a great circle, one whose center coincides with the sphere's. As a dynamical example, the Moon follows a spiral geodesic (in space-*time*) in its orbit about the Earth, the spatial extent of which is about 8×10^5 km (diameter of the Moon's orbit) while the time dimension is about one light-month, or $\sim 8 \times 10^{11}$ km; so that the curvature of the spiral geodesic path is very small, about 1 part in a million; indicative of a weak gravitational field. To find the equation of this geodesic, one moves to a coordinate system freely falling in the gravitational field; writes down the SR force-free equations of motion, $d^2x^i/d\tau^2 = 0$, in a generally covariant form; then changes back to the original coordinate system. The equations of transfer between coordinate systems are, of course, in terms of the metric tensor components which we infer as solutions of the field equations (2.3). We will develop this procedure in detail in Chapter 5.

To summarize: the gravitational dynamics equations of GR are, schematically:

Einstein

space-time curvature $\leftarrow$ mass and energy density

free-fall trajectory $\leftarrow$ geodesic path in curved space-time

---

These are the relativistic versions of Newtonian gravitational dynamics, Equations (1.1) and (1.2), respectively. The first of these is the set of Field Equations of Gravitation, aka the Einstein Field Equations; their derivation is the subject of Chapter 7. The second is the Geodesic Equation derived in Chapter 5.

Problems

1. Derive a formula for the circumference C of a circle of radius r drawn on the surface of a sphere of radius R, and show that $C < 2\pi r$ in all cases. Note that r is the arc length on the surface of the sphere from the center of the circle to its perimeter.
2. Write down the matrix forms of the metric tensors g corresponding to the three standard coordinate systems of Euclidean three-space: Cartesian (x, y, z), cylindrical (ρ, θ, z), and spherical (r, θ, ϕ). Hint: start with the differential line elements as with Equations (2.1) and (2.2).

3

Relativistic cosmology

Cosmology, as the term is currently used, is the study of the structure, contents, and evolution of the Universe on the largest scales. Relativistic cosmological models characterizing the evolution of the Universe on such scales are quantitative forms for the metric tensor components as derived from the Einstein Field Equations. In many cases these are superficially similar to the Newtonian forms of Chapter 1, but they differ from Newtonian models in conceptual interpretations and in several key details. In particular, relativistic cosmological models incorporate all forms of gravitating energy and matter, not just ordinary matter alone; and the character of the models reflects the curvature of space-time in place of Newtonian total energy.

3.1 Cosmological coordinates

The field equations are so complex that their practical solution requires the adoption of a coordinate system fully expressing the symmetries of the application. In cosmology, those symmetries usually arise from the **Cosmological Principle**: on sufficiently large scales the Universe is homogeneous and isotropic, the same everywhere and in all directions (at any given time). The system of galaxies in an expanding, spatially uniform Universe may then be thought of as one in which the galaxies are all falling upward in a uniform gravitational field. This suggests adoption of an appropriately symmetric coordinate system falling with the galaxies, which (by the Equivalence Principle) effectively defines an inertial reference frame, inside which the galaxies are not moving with respect to each other and the physics of SR are valid. Such a coordinate system – which may usefully be visualized as an expanding grid that carries galaxies with it as it expands – is commonly called a **co-moving** coordinate system. Adoption of such a system further suggests the validity of a universal time system, one that applies to all galaxies and that greatly simplifies the physics of universal expansion.

The choice of coordinate system leads to expressions for the metric tensor components in terms of the coordinates. In such a co-moving system the coordinates of a grid point (i.e., galaxy) do not change as a consequence of the expansion of the Universe; the only time-varying element is the **Expansion Function** $a(t)$, which, as in Newtonian cosmology, is non-dimensional and normalized so that $a(t_0) = 1$.

3.2 Expansion and curvature

With the adoption of the co-moving cosmological coordinate system, and consequent metric tensor, the Einstein Field Equations take the form of a simple (but non-linear) differential equation in the expansion function that is superficially very similar to that of Newtonian cosmology (cf. Equation (1.3)):

$$\left(\frac{\dot{a}}{a}\right)^2 = \frac{8\pi G}{3c^2}\varepsilon - \frac{K_0 c^2}{a^2} . \tag{3.1}$$

Here, ε is the total mass–energy density expressed in energy units and K_0 is the geometric curvature of space-time at the current time. Comparing this to Equation (1.3) of Newtonian cosmology: a positive curvature ($K_0 > 0$) corresponds to negative energy ($E < 0$) and a closed expansion history; a negative curvature to positive energy and an open expansion; and $K_0 = E = 0$ to a flat geometry. The resulting forms of the expansion function $a(t)$ are superficially similar to their Newtonian analogs illustrated in Figure 1.1, with subtleties arising from a mixture of gravitating components (matter, radiation) and major alterations arising from dark energy; Part III is devoted to such matters.

All this is true if the Universe contains only matter and radiation. But the possible presence of forces of repulsion, currently attributed to 'dark energy', decouples geometry from expansion so that essentially all currently plausible universal models would expand forever and (eventually) at ever-increasing rates, irrespective of their geometric geometry. This *universal acceleration* was first detected – and has since been further corroborated – by indirect inference of the form of $a(t)$ from observations of distant galaxies. It is a principal objective of modern observational cosmology to determine the form of the expansion function, and thus confirm or deny the existence of dark energy and elucidate the past and future history of universal expansion. A particularly useful tool in such studies is the redshift-distance ('Hubble') relation of distant galaxies.

3.3 Redshifts

Since galaxies in a co-moving coordinate system are not moving with respect to each other, the cosmological redshifts of remote galaxies must arise from some mechanism other than Doppler shifts. That mechanism is the expansion of the co-moving coordinate system, in which everything not bound by some force (internal gravitation, mechanical strength, etc.) expands along with the coordinate system. In particular, photon wavelengths are stretched as the photon travels through space according to[1]

$$\lambda(t) = \lambda(t_0)\; a(t) \;,$$

where t_0 is the current time and $a(t_0) = 1$. Thus, if the photon is emitted from a remote galaxy at time t_e and observed by us at the current time t_0 its redshift z is defined as the fractional wavelength shift $\Delta\lambda/\lambda$:

$$z \equiv \frac{\lambda(t_0) - \lambda(t_e)}{\lambda(t_e)} = \frac{a(t_0) - a(t_e)}{a(t_e)} = \frac{1}{a(t_e)} - 1 \;.$$

The redshift of a galaxy is thus related to the expansion of the Universe by

$$z + 1 = \frac{1}{a} \;, \qquad (3.2)$$

where a is the value of the expansion function at the time photons observed by us were emitted by the galaxy.

In essence, galaxy redshifts map out not the expansion *rate* of the Universe as in classical cosmology, but the expansion *history*: light from a $z = 1$ galaxy, for instance, was emitted at a time when the Universe was half its present size. There is no mention of velocities in this interpretation, nor (directly) of expansion rate since time does not appear explicitly and is not directly observable. To use this relation to infer the form of $a(t)$ we need another connection between redshifts and observable properties of distant galaxies. The most commonly employed such relation is that involving galaxy distance: the Hubble Relation.

3.4 Hubble Relation

In classical (non-GR) analyses of universal expansion, radial velocities of distant galaxies are inferred from spectral shifts interpreted as Doppler shifts:

$$z \equiv \frac{\lambda_{\text{observed}} - \lambda_{\text{emitted}}}{\lambda_{\text{emitted}}} \approx \frac{V}{c}$$

(for $V \ll c$). In terms of redshift the Newtonian Hubble relation $V = Hd$ (I.2) then takes its modern form of distance proportional to redshift:

$$d = \frac{V}{H} \approx \frac{c}{H} z \qquad (z \ll 1) \;, \qquad (3.3)$$

[1] This relation is proven rigorously in Chapter 9.

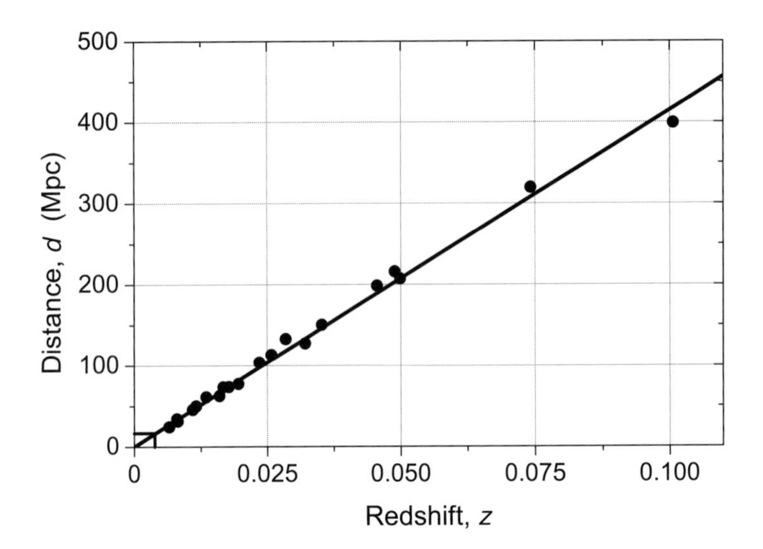

Figure 3.1 Hubble Relation for low-redshift galaxies beyond the Local Group (deduced from SN Ia supernovae in nearby galaxies and clusters). The box in the lower left-hand corner encompasses the range of Figure 4 of the Introduction.

where $H = \dot{a}/a$. Since both d and z for galaxies are observable (with some qualifications), one can use this relation to calibrate the value of the expansion rate H from nearby galaxies. Figure 3.1 demonstrates the quality of this linear fit to modern data for relatively nearby galaxies.

When applied to the wider Universe beyond nearby galaxies, however, this tidy result is beset with conceptual difficulties. First, the meaning of 'distance' is confused by the conflation of distance with time in an expanding Universe: does the d used in these equations apply to the distance *now*, or to that when the photons we now observe were emitted? On cosmological scales these two distances may differ considerably. Second, in a Universe dominated by gravitation we expect the expansion to gradually slow down so that H should decrease with time; in effect, that the proper value for H should differ from one galaxy to the next (if they lie at different distances). Clearly, the Hubble Relation of Equation (3.3) needs to be thought through a bit more carefully if it is to be used to study the Universe's expansion history.

The resolution of these ambiguities is a general relativistic one as derived in Part III of this book, and we will leave further development of the relation until then. For the nearest galaxies, however, we can presume that $t \approx t_0 \Rightarrow H \approx H_0$ (its current value), and d is given by the classical d^{-2} dependence of luminous flux for calibrating galaxies. Then, to first order,

$$\lim_{z \to 0} d(z) = \frac{c}{H_0} z \,, \tag{3.4}$$

from which the current expansion rate, H_0, may be inferred from observed distances and redshifts of nearby galaxies, much as Hubble did in his original work in the 1920s. Applied to nearby galaxies this relation implies the currently estimated value $H_0 \approx 0.074$ Gyr^{-1} (or, in the formulation of Equation (I.2), ≈ 72 km/sec/Mpc).

3.5 Expansion history

The expansion history of the Universe is complicated by the existence of three sources of gravitation: matter (m), radiation (r), and dark energy (de). The densities of the these sources vary as the Universe expands, and they do so in different manners so that the dominant source of gravitation changes as the Universe expands, and so then does its expansion rate.

3.5.1 Energy densities

It is convenient to represent matter density as an energy density, in the usual SR manner: $\varepsilon_{\mathrm{m}} = \rho c^2$. Then, in the absence of matter creation, the energy density of matter scales simply as $\varepsilon_{\mathrm{m}}(a) \propto a^{-3}$ as co-moving volumes expand with the Universe. With radiation, however, there is an additional energy diminution caused by redshift, of the form $\varepsilon_{\mathrm{photon}} = hc/\lambda_{\mathrm{photon}} \propto a^{-1}$; so radiation energy density varies as $\varepsilon_{\mathrm{r}}(a) \propto a^{-4}$. The true nature of dark energy is still much of a mystery, but if it arises from a 'Cosmological Constant' as currently believed by many, its energy density does not change as the Universe evolves. The consequence of this mixture is that as the Universe expands, different energy sources become dominant in its gravitation: radiation in the earliest phase, dark energy in the latest, and matter (perhaps) at intermediate times.

The current energy densities are believed to be in the approximate ratios

$$\varepsilon_{\mathrm{r}} : \varepsilon_{\mathrm{m}} : \varepsilon_{\mathrm{de}} \quad \approx \quad 10^{-4} : 0.3 : 0.7\,,$$

so dark energy currently dominates the expansion, but not by much; and radiation is an insignificant contributor. But at sufficiently early times – those for which $a \lesssim 10^{-4}$ ($\Rightarrow z \gtrsim 10^4$) – radiation was the dominant source of gravitation. During this early era of radiation dominance the Universe expanded more slowly than it would otherwise have done, as expansion energy was invested in redshifting photons. In the current era, dark energy dominates universal expansion, which is nearly exponential (as you will be invited to show in a problem).

3.5.2 Expansion eras

Sufficiently early in the Universe's expansion, when particle densities were very high, particle scattering rates would have been so high that everything would have been in thermal equilibrium and radiation would have been that of a black body: $\varepsilon_r = a_{rad}T^4$, where a_{rad} is the black-body energy density constant (see Appendix C). Since the energy density of radiation evolves as $\varepsilon_r \propto a^{-4}$, the temperature of black-body radiation, and consequent mean particle energies, vary as a^{-1} and are therefore arbitrarily large at early times. This allows a conceptualized history of the Universe as parameterized by its temperature or, equivalently, mean particle energy. This history is usefully divided into eras and events as follows.

Particle Era Prior to $t \approx 1$ second the temperatures are so high that mean photon energies exceed the rest energies of many fundamental particles and all the action is in terms of particle physics and quantum fields. The details of this era depend on how far back one wishes to go: prior to $t \sim 10^{-5}$ seconds the physics is poorly known; ultimately ($t \sim 10^{-35}$ seconds) we reach the era of quantum exotica that is more the realm of quantum speculation than of astronomy, and that includes the hypothetical (but widely believed) inflation of the Universe by many orders of magnitude.

Eventually the expanding Universe cools to the point that compound nuclei can survive and primordial nucleosynthesis occurs between $t \sim 1$ and ~ 10 minutes, when the plasma is sufficiently cool but still hot enough for fusion to take place. In this brief period of time nearly all the Universe's helium forms, along with trace amounts of other light isotopes (lithium, deuterium); but by the time helium has formed in any quantity the temperature has fallen below the threshold for fusion of heavier elements so that essentially little beyond helium forms in the Big Bang.

Plasma Era Thereafter the Universe is a hot, expanding, cooling plasma of protons, helium nuclei, miscellaneous other light nuclei, electrons, and radiation. Scattering of photons off free electrons during this era keeps the matter in thermal equilibrium with the radiation, so they share a common temperature. As this plasma expands the density of radiation energy decreases more rapidly $\left(\propto a^{-4}\right)$ than does that of matter $\left(\propto a^{-3}\right)$ until, at an age of $\sim 50,000$ years, matter takes over as the dominant source of gravitation and the Universe's expansion speeds up a bit. But the temperature still remains too high for neutral atoms to exist.

At the age of $t \sim 380,000$ years the temperature has fallen to the point ($T \sim 3000$ K) that electrons can be captured onto positive nuclei, and neutral atoms (mostly H and He) appear. This is the time of **photon decoupling**, often called **recombination,** an unfortunate misnomer since there has been no previous combination for this to be the **re**- of. As free electrons disappear, photon scattering off matter largely ceases and the Universe becomes transparent for the first time in its history. The remnant black-body radiation from the Plasma Era is free to expand into the Universe where it cools as before, but this time independently of matter. The time of photon decoupling from matter corresponds to $a \approx 9 \times 10^{-4}$, or $z \approx 1100$. Thus, by the current time this **Cosmological Microwave Background Radiation (CMB)** has cooled to $T_{\text{CMB}} \approx 3000/1100 \approx 2.7$ K, where it is observed as a ubiquitous microwave background coming from all directions. This – the age of photon decoupling or of last scattering – is the earliest phase of the Universe that is directly observable (earlier phases must be indirectly inferred by, e.g., the abundance of helium, or anisotropies in the CMB).

Era of Galaxies During the Plasma Era radiation pressure essentially damped out any structure formation from gravitational clumping of ordinary matter. Once radiation has decoupled from it, matter is free to form gravitational structures in the forms of galaxies and clusters of galaxies; and starlight can act to re-ionize the intergalactic medium. The relatively short period of electrical neutrality is called the Dark Ages, which precedes the **Era of Galaxies**. During this era – lasting for at least the last 13 Gyr – matter and, later, dark energy have dominated the Universe's gravitating content, and radiation has played an insignificant role.

The beginning of the Galaxy Era, when stars first form, marks the end of the relative simplicity of the Universe as radiative and mechanical feedback from the first stars severely complicates the physics of matter in an expanding Universe. Thereafter galaxies and galaxy clusters form by processes whose details are the subject of much current research; and eventually produce the Universe that we currently perceive.

Problems

1. Show that a solution to the relativistic expansion equation (3.1) for a flat geometry and a constant energy density ε_0 is exponential expansion, and compute an estimate for the resulting Hubble Parameter $H = \dot{a}/a$ if the energy density is $\varepsilon_0 = 10^{-10}$ J/m^3. Compare your results to those of Problem 2 in Chapter 1.
2. Estimate the Hubble Constant implied by the data in Figure 3.1, in units of both km/sec/Mpc and Gyr^{-1}.
3. From the current energy densities listed in Section 3.5.1 estimate the value of the expansion function at the time when radiation and matter energy densities were the same. What was the dark energy density at that time?

Part II

General Relativity

4

General covariance

A major accomplishment of Einstein's Theory of Special Relativity (SR) was the demonstration that the laws of physics took the same form for all inertial observers. In more relevant language we would say that the laws of physics were independent of the reference frame and coordinate system chosen for their expression – but only for inertial reference frames or, what is equivalent, for coordinate systems related by Lorentz transforms. This *covariance* of inertial coordinate systems allowed Einstein to write the laws of mechanics and of electrodynamics in ways that revealed new aspects (e.g., $E = mc^2$) and extended their validity to reference frames moving at high velocity.

It was thus a principal purpose of General Relativity (GR) to extend this **Principle of General Covariance** – that the equations of physics were invariant to change of coordinate systems – to all reference frames, including accelerating ones. Needed for this purpose are tools of mathematical physics that preserve equalities under *all* changes of coordinate systems and thus – as Einstein would put it – free physics from the tyranny of coordinates. The required tools were providentially at hand in the form of the Absolute Differential Calculus, a coordinate-independent version of calculus being developed by (mostly Italian) mathematicians. Among the tools in that development were mathematical quantities associated with geometry that all transformed in the same manner under changes in coordinate systems, so that equalities in any coordinate system led to equalities in all. These were **tensors**, different forms for which have since been employed in many areas of advanced physics.

The idea of tensors is simple enough: if tensors A and B are equal in coordinate system S, and transform to A' and B' in system S', then it follows that $A' = B'$ there also. The trick is to find tensors to represent physical quantities of interest; fortunately for GR, in which geometrically related objects play a central role, tensor representations exist for most applications.

The remainder of this chapter is devoted to an explanation of the most basic properties of tensors and of their manipulations in the service of the differential geometry employed in GR. Tensor analysis is not really difficult, but it *is* different from the mathematics to which students have normally been exposed and so requires one's attention in order to understand. The level of tensor analysis used in basic GR is fairly straightforward, and we have endeavored to keep it as simple as possible in what follows. Just read the rest of the chapter carefully and work through the problems and you should have no trouble understanding the mathematics of GR to come.

4.1 Tensors

The most accessible representation of tensors is that of indexed arrays of mathematical/physical quantities. The indexing stems from that of the coordinate system employed: because the mathematical relations encountered in GR are typically very complex – often encompassing hundreds or thousands of terms – an indexed coordinate system is needed for comprehension. Thus, we represent coordinate systems with a generic symbol such as x, and individual coordinates as superscripted instances, such as x^i. For instance, in Euclidean 3-space we represent Cartesian and polar coordinate systems as

$$\left(x^1, x^2, x^3\right) = \begin{cases} (x, y, z) \\ (r, \theta, \phi) \end{cases} ,$$

and so forth.

4.1.1 Coordinate systems

The change of coordinates from system $\{x\}$ (e.g., polar) to system $\{\bar{x}\}$ (e.g., Cartesian) entails transformation equations that we generically write as $\bar{x}^i = \bar{x}^i(\{x\})$ so that, for the above coordinate systems, $y = r \sin\theta \sin\phi$, or $\bar{x}^2 = x^1 \sin x^2 \sin x^3$; and so forth. Such transformation equations will always be differentiable (in our applications), so for this example we have

$$\begin{aligned} dy &= \frac{\partial y}{\partial r} dr + \frac{\partial y}{\partial \theta} d\theta + \frac{\partial y}{\partial \phi} d\phi , \\ &= \sin\theta \sin\phi \, dr + r \cos\theta \sin\phi \, d\theta + r \sin\theta \cos\phi \, d\phi . \end{aligned}$$

Or, in indexed coordinates,

$$d\bar{x}^2 = \sin x^2 \sin x^3 \, dx^1 + x^1 \cos x^2 \sin x^3 \, dx^2 + x^1 \sin x^2 \cos x^3 \, dx^3 .$$

We can generically write such things as

$$d\bar{x}^i = \sum_j \frac{\partial \bar{x}^i}{\partial x^j} dx^j \,. \tag{4.1}$$

The compactness of this result is a consequence of the use of generalized superscripted coordinate symbols. This compaction of systems of equations will be encountered throughout GR; make sure you understand it.

We will shortly be dealing with more complicated relations involving nested summations, and it is useful to adopt a notation that avoids the clumsiness of such expressions. You will note in the above equation that the summation occurs over an index (j) that is repeated. This is so common an occurrence that we adopt the following **summation convention** to eliminate the summation sign:

> *A summation is implied over any index that occurs twice in an expression, once as a superscript and once as a subscript.*

For this to apply to Equation (4.1) we treat differentiation $\partial/\partial x^i$ as a subscripted operator. Indeed, in many forms of mathematical physics this is written as ∂_i $\left(\equiv \partial/\partial x^i\right)$ and thus is explicitly subscripted. With this convention we can write equations such as (4.1) in the form

$$d\bar{x}^i = \frac{\partial \bar{x}^i}{\partial x^j} dx^j \,, \tag{4.2}$$

with the understanding that a summation over j is required.

By convention, in four-dimensional space-time the 0 index is used for time, and indices 1 through 3 for spatial dimensions (as above). Notationally, Latin indices $(i, j, \ldots)$ denote spatial coordinates $(1, 2, 3)$, while Greek indices $(\mu, \nu, \ldots)$ denote all four space-time coordinates $(0, 1, 2, 3)$. For the purposes of tensor analysis *per se*, either form of indices will suffice, so we shall typically use Latin indices throughout this chapter.

4.1.2 Tensor types

Tensors come in two types, **covariant** and **contravariant**, and in different dimensions or **ranks**. The simplest non-trivial tensors are of rank 1 and take the mathematical form of vectors. Their transformation properties upon change of coordinate systems are:

$$\text{covariant:} \quad \bar{V}_i = V_j \frac{\partial x^j}{\partial \bar{x}^i}, \tag{4.3}$$

$$\text{contravariant:} \quad \bar{V}^i = V^j \frac{\partial \bar{x}^i}{\partial x^j}. \tag{4.4}$$

By convention, covariant tensors carry subscripts and contravariant ones are superscripted. Note that in both of the above equations a summation over j is implied, but the repeated index (on the right) occurs in somewhat different forms. Note also from Equation (4.2) that coordinate differentials themselves are contravariant tensors.

While rank-1 tensors take the form of vectors, *not all vectors are tensors!* The tensor characteristic lies in the form of transformation upon change of coordinates, and is responsible for the invariance of tensor equations. If we have, say, equality between covariant, rank-1 tensors $\bar{A}$ and $\bar{B}$ in the $\{\bar{x}\}$ coordinate system, so that $\bar{A}_i = \bar{B}_i$ for all indices i; and then change to system $\{x\}$ so that $\bar{A} \to A$ and similarly for $\bar{B}$, the covariant transformation equations (4.3) imply

$$
\begin{aligned}
A_i \frac{\partial x^i}{\partial \bar{x}_j} &= B_i \frac{\partial x^i}{\partial \bar{x}_j}\,, \\
\Rightarrow (A_i - B_i)\frac{\partial x^i}{\partial \bar{x}_j} &= 0.
\end{aligned}
\tag{4.5}
$$

This last equation carries an implicit sum over i and so is of the form $V \cdot \mathbb{M} = 0$ where V is the vector $A - B$ and $\mathbb{M}$ is the Jacobian matrix $[\partial x / \partial \bar{x}]$. For the change of coordinate systems to be a valid one this matrix must be non-singular so that the only solution to Equation (4.5) is $A = B$. *Tensor equations are invariant to changes of coordinate systems.* While we have proved this only for one particular type of tensor, the proofs for other types and ranks follow easily. *A valid tensor equation in one coordinate system is equally valid in all others.*

The transformation characteristics of tensors typically arise from geometrically related invariants, as illustrated in the following example.

Example: a covariant tensor

Consider a scalar function S of the coordinates $\{x\}$ that describes a surface in multidimensional space:

$$S(\{x\}) = C\,,$$

for some constant C. An example would be the surface of a sphere of radius R centered at the origin: in Cartesian coordinates, $S\left(x^1, x^2 x^3\right) = \left(x^1\right)^2 + \left(x^2\right)^2 + \left(x^3\right)^2 = R^2$. The normal vectors to this surface are

$$\vec{N} = \vec{\nabla} S \quad \Rightarrow \quad N_j = \frac{dS}{dx^j} = 2x^j\,, \tag{4.6}$$

for $j = 1, 2, 3$. Such normal *vectors* $\vec{N}$ have well-defined magnitudes and directions that depend only upon the surface and are independent of the coordinate system chosen, but their *components* (such as N_j) *will* change with the coordinate system. In the $\bar{x}$ system they will be

$$\bar{N}_i = \frac{dS}{d\bar{x}^i} . \tag{4.7}$$

Now, by the chain rule for derivatives,

$$\frac{dS}{d\bar{x}^i} = \frac{dS}{dx^j}\frac{\partial x^j}{\partial \bar{x}^i} \quad \text{(implied sum on } j) .$$

Comparing these last two equations to Equation (4.6),

$$\Longrightarrow \bar{N}_i = \frac{dS}{dx^j}\frac{\partial x^j}{\partial \bar{x}^i} = N_j \frac{\partial x^j}{\partial \bar{x}^i} ,$$

from which we see that the components of normal vectors to a surface are covariant, a property that arises from the invariant nature of the vectors themselves, which represent quantities independent of coordinates.

A similar demonstration of simple contravariant vectors is provided in Problem 5 at the end of this chapter. These examples illustrate a subtle feature of tensors: tensors themselves (e.g., geometrical objects) are independent of coordinates, but their *components* are usually not. Strictly speaking, we should say that tensors are *invariant* to changes of coordinates, while their components are either covariant or contravariant (or neither). But the terms *covariant* and *contravariant* are commonly applied to the tensors themselves, usually without confusion.

4.1.3 Higher rank tensors

Tensors of rank higher than 1 are similarly defined. Rank-2 tensors, the type we will mostly be dealing with here, have two indices and take the mathematical form of matrices. The covariant and contravariant forms transform as

$$\bar{A}_{ij} = A_{mn}\frac{\partial x^m}{\partial \bar{x}^i}\frac{\partial x^n}{\partial \bar{x}^j} , \tag{4.8}$$

$$\bar{A}^{ij} = A^{mn}\frac{\partial \bar{x}^i}{\partial x^m}\frac{\partial \bar{x}^j}{\partial x^n} . \tag{4.9}$$

(Note: implied nested sums over m and n.) Needless to say, not all matrices represent tensors and transform this way. Those that do arise from invariants with respect to transformation of coordinate systems. A prime example is the metric tensor discussed in the following section.

Transformation rules for tensors of rank higher than 2 are trivial extensions of those for rank-2 tensors (Equations (4.8) and (4.9)); e.g.,

$$\bar{T}^{ijk} = T^{mnp} \frac{\partial \bar{x}^i}{\partial x^m} \frac{\partial \bar{x}^j}{\partial x^n} \frac{\partial \bar{x}^k}{\partial x^p} ,$$

and similarly for the covariant form. Tensors of rank higher than 1 can also exist in mixed form; e.g.,

$$\bar{A}^i_j = A^m_n \frac{\partial \bar{x}^i}{\partial x^m} \frac{\partial x^n}{\partial \bar{x}^j} .$$

You might want to study this to ensure that the indices are properly placed. A more complicated example that will be of some importance later in this book is the Riemann Curvature Tensor, a mixed, rank-4 tensor that transforms as

$$\bar{\mathcal{R}}^{\lambda}{}_{\mu\nu\kappa} = \mathcal{R}^{\alpha}{}_{\beta\gamma\sigma} \frac{\partial \bar{x}^\lambda}{\partial x^\alpha} \frac{\partial x^\beta}{\partial \bar{x}^\mu} \frac{\partial x^\gamma}{\partial \bar{x}^\nu} \frac{\partial x^\sigma}{\partial \bar{x}^\kappa} .$$

Finally, as a generic template for tensors of whatever form and rank,

$$\bar{R}^{ab\cdots c}_{de\cdots f} = R^{\alpha\beta\cdots\gamma}_{\delta\varepsilon\cdots\phi} \left[\frac{\partial x^\delta}{\partial \bar{x}^d} \frac{\partial x^\varepsilon}{\partial \bar{x}^e} \cdots \frac{\partial x^\phi}{\partial \bar{x}^f} \right] \left[\frac{\partial \bar{x}^a}{\partial x^\alpha} \frac{\partial \bar{x}^b}{\partial x^\beta} \cdots \frac{\partial \bar{x}^c}{\partial x^\gamma} \right] .$$

These look worse than they actually are: if you study them for a moment you will see that there is only one way to assign indices that makes sense (and for which the summation convention works).

4.2 Metric tensor (II)

As introduced in Chapter 2, the invariant space-time interval ds is given in terms of coordinate differentials by (Equation (2.2))

$$ds^2 = g_{ij}\, dx^i dx^j , \tag{4.10}$$

where g_{ij} are components of the **metric tensor g**. Since ds is the same in all coordinate systems (distances are invariant under changes of coordinates), it must be that this same equation holds in a new coordinate system denoted by, say, $\{\bar{x}\}$:

$$ds^2 = g_{ij}\, dx^i dx^j = \bar{g}_{mn} d\bar{x}^m d\bar{x}^n , \tag{4.11}$$

where $\bar{g}_{mn}$ are components of the metric tensor $\bar{\mathbf{g}}$ in the new coordinate system (note: implied double sums). From Equation (4.2)

$$d\bar{x}^m = \frac{\partial \bar{x}^m}{\partial x^i} dx^i$$

for each $\bar{x}^m$. Doing the same for $\bar{x}^n$ yields

$$\bar{g}_{mn} d\bar{x}^m d\bar{x}^n = \left[\bar{g}_{mn} \frac{\partial \bar{x}^m}{\partial x^i} \frac{\partial \bar{x}^n}{\partial x^j} \right] dx^i dx^j \,.$$

Inserting this into the right-hand side of Equation (4.11) and re-arranging,

$$\left[g_{ij} - \bar{g}_{mn} \frac{\partial \bar{x}^m}{\partial x^i} \frac{\partial \bar{x}^n}{\partial x^j} \right] dx^i dx^j = 0 \,.$$

Since the coordinates in each system are linearly independent, this equality can hold only if the quantity in brackets is identically zero, so that

$$g_{ij} = \bar{g}_{mn} \frac{\partial \bar{x}^m}{\partial x^i} \frac{\partial \bar{x}^n}{\partial x^j} \,, \tag{4.12}$$

which is the transformation rule for rank-2, covariant tensors. The metric tensor is, indeed, a tensor (covariant, rank-2). Its tensor nature is a consequence of the invariance of the space-time interval in relativity.

Since (like all rank-2 tensors) the metric tensor $\mathbf{g}$ takes the mathematical form of a matrix, which must be non-singular in any meaningful physical application, it possesses a matrix inverse $\mathbf{g}^{-1}$ with components denoted by g^{ij}, so that

$$g_{ik} g^{kj} = \mathbb{I}_i^j = \delta_i^j \,, \tag{4.13}$$

where $\mathbb{I}$ is the identity matrix or, equivalently, δ is the Kronecker delta:

$$\mathbb{I}_i^j = \delta_i^j = \begin{cases} 1\,, & i = j, \\ 0\,, & i \neq j. \end{cases} \tag{4.14}$$

Note that $\mathbb{I}$ is an invariant tensor (its components are the same in all coordinate systems); this fact, combined with Equation (4.13), implies that the metric tensor inverse is also a tensor: contravariant, rank-2. For diagonal metrics (the only types we shall be using), $g^{ij} = 1/g_{ij}$ and the inverse metric is easily computed without the need of a full matrix inverse calculation.

4.3 Tensor manipulations

Any direct product of tensors, and any linear combination of tensors of the same type and rank, is also a tensor, as can easily be verified by examining the transformation properties. Thus:

$$\begin{aligned} C_{ij} &= A_i B_j \,, \\ D_i^j &= A_i B^j \,, \\ E^i &= aA^i + bB^i \,. \end{aligned}$$

(where a and b are scalar constants) are all tensors. We can use these relations to change the type of a tensor by, e.g.,

$$C_i = B_{ij}A^j . \tag{4.15}$$

Examine this carefully to see what has happened here: because a sum on j is implied, the contravariant A has effectively been changed to the covariant C. This turns out to be a crucial tool in developing the field equations of gravitation.

4.3.1 Rank reduction

We can reduce the rank of a mixed-rank tensor by setting an upper index equal to a lower and performing the implied sum:

$$R_{jk} = R^i_{\ jki} \qquad \left(= \sum\nolimits_i R^i_{\ jki}\right) . \tag{4.16}$$

If the original mixed-rank tensor was a true tensor, so will be its reduced form. A somewhat extreme form of this manipulation is the creation of a rank 0 tensor (i.e., a scalar) from a mixed-rank-2 tensor:

$$F = F^i_i .$$

Note that this sort of rank reduction is really a special kind of tensor manipulation, for seemingly similar constructs are not necessarily tensors. Thus:

$$S_{jk} = \sum_i S_{ijki} ,$$

$$A_i = A_{i3} ,$$

are probably not tensors at all, or at least not useful ones. As always, the putative tensor nature of an expression can be examined via its transformation equations, but not always easily.

4.3.2 Changing tensor types

Tensors can be changed from covariant to contravariant, and vice versa, by manipulations with other tensors, such as illustrated by Equation (4.15). In applications to differential geometry this operation is particularly useful when the 'index raising/lowering' tensor is the metric tensor g, whereupon we normally use the same symbol for both the original and the transformed tensor. Thus:

$$A_i = g_{ij}A^j ,$$

$$B^m = g^{mn}B_n .$$

Tensor type changing can be combined with rank reduction to produce almost any desired tensor type associated with a given tensor. As examples:

$$R^i_{kl} = g^{ij} R_{jkl} ,$$
$$T^j_{\ k} = g_{ik} T^{ij} .$$

Rank reduction to a scalar can be envisioned as a two-step process; e.g., (1) $T^i_j = g_{jk} T^{ik}$, followed by (2) equating upper and lower indices and summing: $T = T^i_i$. This is clearly equivalent to $T = g_{ik} T^{ik}$, which is simpler and easier to remember.

4.4 Derivatives

To finish off a rather abstract chapter, here is an instructive example of a familiar mathematical object that is *not* a tensor: the ordinary (i.e., coordinate) derivative of a tensor. Starting from the transformation rule for a contravariant tensor,

$$B^\mu = \frac{\partial x^\mu}{\partial \bar{x}^\nu} \bar{B}^\nu ,$$

and differentiating both sides with respect to x^κ, we have

$$\begin{aligned} \frac{\partial B^\mu}{\partial x^\kappa} &= \frac{\partial x^\mu}{\partial \bar{x}^\nu} \frac{\partial \bar{B}^\nu}{\partial x^\kappa} + \frac{\partial^2 x^\mu}{\partial x^\kappa \partial \bar{x}^\nu} \bar{B}^\nu , \\ &= \overbrace{\frac{\partial x^\mu}{\partial \bar{x}^\nu} \frac{\partial \bar{x}^\lambda}{\partial x^\kappa} \frac{\partial \bar{B}^\nu}{\partial \bar{x}^\lambda}} + \frac{\partial^2 x^\mu}{\partial x^\kappa \partial \bar{x}^\nu} \bar{B}^\nu , \end{aligned} \tag{4.17}$$

where in the last step we have used the identity

$$\frac{\partial}{\partial x^\kappa} = \frac{\partial \bar{x}^\lambda}{\partial x^\kappa} \frac{\partial}{\partial \bar{x}^\lambda}$$

in order to effect the differentiation of $\bar{B}$ with respect to $\bar{x}$, instead of x. The term with the overbrace is exactly what would be expected if $\partial B^\mu / \partial x^\kappa$ were a mixed, rank-2 tensor; the additional term with the second derivative spoils the tensor nature of the derivative. *Ordinary (coordinate) derivatives of tensor components are not tensors.*

The origin of the difficulty lies in the non-local nature of differentiation. In general,

$$\frac{\partial V^\mu}{\partial x^\kappa} = \lim_{\delta x^\kappa \to 0} \left[\frac{V^\mu (x^\kappa + \delta x^\kappa) - V^\mu (x^\kappa)}{\delta x^\kappa} \right] .$$

But $V^\mu (x^\kappa + \delta x^\kappa)$ and $V^\mu (x^\kappa)$ are evaluated at two different places so that their difference reflects not only the change in V but also the change in orientation of the coordinate axes. Such a change can arise from coordinate systems in flat geometries that change direction with position (e.g., polar coordinates), or from curvature of the underlying space (e.g., on the surface of a sphere). In either case the coordinate derivative consists of the 'true' change in the tensor, plus an extraneous term that reflects the coordinate system. To put this into perspective, ordinary derivatives are not generally covariant because they do not represent physical quantities that are invariant with respect to changes of coordinate systems.

This is a most unfortunate situation that differential geometers rectify by re-defining differentiation so that it (1) produces generally covariant results when applied to tensors, and (2) reduces to the ordinary definition when the underlying space is flat and the coordinates Cartesian. This is a **covariant derivative**, discussed in some detail in Appendix A, Section A.3. It takes the form of the ordinary derivative plus a correction term that removes the effect of coordinate system changes from the derivative, leaving the 'pure' change in the tensor as its derivative.[1] The covariant derivative operator may be applied to tensors of any rank or type.

Note that, from Equation (4.17), we expect differentiation to append a covariant index to the tensor being differentiated, hence the name *covariant* derivative. This also justifies our previously stated interpretation of the derivative operator, $\partial/\partial x^{\mu}$, as a subscripted quantity for the purposes of the Einstein summation convention.

The notation for derivatives varies from one writer to another. The most commonly encountered are:

$$\begin{aligned} \text{ordinary derivative:} \quad & \frac{\partial}{\partial x^{\mu}} A^{\cdots}_{\cdots} \rightarrow A^{\cdots}_{\cdots,\mu}, \\ \text{covariant derivative:} \quad & D_{\mu} A^{\cdots}_{\cdots} \quad \rightarrow A^{\cdots}_{\cdots;\mu}. \end{aligned}$$

That is, a (lower index) comma for ordinary derivatives and a (covariant) semicolon for covariant derivatives. Thus $R_{jk,m}$ and $R_{lj;m}$ are, respectively, the ordinary and covariant derivatives of R_{jk} (and similarly for different tensor types and ranks).

Problems

1. Complete the following:

$$\begin{aligned} g_{\mu\nu} A^{\mu}_{\kappa} &= A^{?}_{?}, \\ B^{\alpha\eta\delta}_{\beta\delta} &= B^{?}_{?}, \end{aligned}$$

[1] This is the origin of the complicated expressions for, e.g., divergence in polar and cylindrical coordinates.

$$
\begin{aligned}
g_{\mu\nu} g^{\kappa\mu} &= ?, \\
g_{\sigma\kappa} g^{\kappa\tau} C_{\tau\mu} &= C^?_?, \\
D^?_? E^{\alpha\beta} &= F^{\alpha}_{\ \gamma}, \\
T^{\rho\tau} g^{\kappa}_{\rho} &= T^?_?, \\
\bar{A}^{jk}_{i} &= A^{ab}_{\ \ c} \frac{\partial ?}{\partial ?} \frac{\partial ?}{\partial ?} \frac{\partial ?}{\partial ?}.
\end{aligned}
$$

2. Use the metric tensor (and its inverse), and/or rank contraction, to change the rank-1 contravariant, rank-2 covariant tensor $S^{\alpha}_{\ \beta\gamma}$ to (1) rank-2 contravariant, rank-1 covariant, $S^{\cdot\cdot}_{\cdot}$; (2) rank-3 covariant, $S_{\dots}$; (3) rank-1 contravariant, $S^{\cdot}$; (4) rank-1 covariant, $S_{\cdot}$.
3. Numerically evaluate $g_{\mu\nu} g^{\mu\nu}$ for a four-dimensional space-time metric.
4. Consider a two-dimensional space with covariant metric tensor

$$
\boldsymbol{g}_{\cdot\cdot} = \begin{pmatrix} A & 0 \\ 0 & B \end{pmatrix},
$$

and a rank-2 covariant tensor

$$
T_{\cdot\cdot} = \begin{pmatrix} C & 1 \\ -1 & 0 \end{pmatrix}.
$$

Display the matrix forms of $T^{\cdot\cdot}$ and $T^{\cdot}_{\cdot}$, and the scalar T .

5. The tangent vectors $\tilde{\mathbf{V}}$ along a space curve given by $\vec{\mathbf{x}} = \left\{x^i(s)\right\}$ for some parameter s that increases along the curve (e.g., the cumulative arc length) have components $V^i = dx^i/ds$. Show that the V^i are contravariant, a consequence of the invariance of tangent vectors with respect to changes of coordinates.

5

Equivalence Principle

General covariance has something of the character of a *strategy*, rather than as a fundamental principle. It probably would be possible to construct a theory of GR without resorting to tensors or other forms of general covariance, although it certainly would not be easy to do so. Equivalence, on the other hand, is absolutely fundamental to GR, constituting the connection between curvature and gravitation. As a principle underlying GR it has its basis in the observed equality of gravitational and inertial mass, so that all objects experience the same acceleration in a gravitational field, quite unlike the case with, e.g., electrical or magnetic fields. By itself, this sets gravitation apart from other forces; but Einstein extended the principle further by applying it – in some sense – to all of physics, not just gravitational dynamics. It is thus useful to distinguish between the **Weak Equivalence Principle** and the **Strong Equivalence Principle,** both of which assert that acceleration and gravitation are equivalent, but in somewhat different senses and with different consequences.

5.1 Weak Equivalence Principle

The Weak Equivalence Principle (WEP) states simply that all objects in a gravitational field experience the same acceleration. As a consequence, the accelerating effects of gravitation can be transformed away by going over to a coordinate system falling freely with the gravitational field, much like astronauts in the space station or Einstein's workmen falling from a roof. Thus, an operational definition of the WEP is:

> The dynamical effects of a gravitational field can be transformed away by moving to a reference frame that is freely falling in the gravitational field.

The utility of this principle is limited by its strict applicability only to static, uniform gravitational fields, which are rare. Objects falling to the ground, for instance, experience a gravitational force that increases as they get closer to the Earth; and objects in orbit experience a gravitational field that changes direction with orbital location. Applications of the WEP are thus limited to the near locality (in space-time) of a falling object: such **Locally Inertial Reference Frames** are formally defined in the following section.

Even with this limitation the WEP can have interesting consequences for gravitational physics, the most important of them being the equations of motion of a freely falling object – i.e., one experiencing only gravitational forces.

5.1.1 Gravitational Equations of Motion

Consider an object falling freely in an arbitrary gravitational field. In a reference frame falling with the object there will be no accelerations and the dynamics will be those of SR. Specifically, the equations of motion of the object will be

$$\frac{d^2\xi^\alpha}{d\tau^2} = 0\,, \tag{5.1}$$

where, following tradition and convention, ξ denotes Cartesian, non-accelerating coordinates (within the freely falling reference frame) and τ is the SR proper time. We now switch to an arbitrary coordinate system x^μ, in terms of which $\xi^\alpha = \xi^\alpha(\{x^\mu\})$; and re-write the equations of motion in this system:

$$\begin{aligned} 0 = \frac{d^2\xi^\alpha}{d\tau^2} &= \frac{d}{d\tau}\left(\frac{d\xi^\alpha}{d\tau}\right)\,, \\ &= \frac{d}{d\tau}\left(\frac{\partial\xi^\alpha}{\partial x^\nu}\frac{dx^\nu}{d\tau}\right)\,, \end{aligned} \tag{5.2}$$

where we have employed the usual chain rule of coordinate derivatives,

$$d\xi^\alpha = \frac{\partial\xi^\alpha}{\partial x^\nu}dx^\nu\,. \tag{5.3}$$

Carrying the time derivative through the expression in parentheses in Equation (5.2),

$$0 = \frac{\partial\xi^\alpha}{\partial x^\nu}\frac{d^2x^\nu}{d\tau^2} + \frac{dx^\nu}{d\tau}\frac{d}{d\tau}\frac{\partial\xi^\alpha}{\partial x^\nu}\,. \tag{5.4}$$

Again invoking the chain rule for derivatives, this time as

$$\frac{d}{d\tau} = \frac{dx^\mu}{d\tau}\frac{\partial}{\partial x^\mu}\,,$$

so that

$$\frac{d}{d\tau}\frac{\partial \xi^\alpha}{\partial x^\nu} = \frac{dx^\mu}{d\tau}\frac{\partial^2 \xi^\alpha}{\partial x^\nu \partial x^\mu} .$$

Substituting this into Equation (5.4) gives an alternative expression for the equations of motion:

$$\frac{\partial \xi^\alpha}{\partial x^\nu}\frac{d^2 x^\nu}{d\tau^2} + \frac{\partial^2 \xi^\alpha}{\partial x^\nu \partial x^\mu}\frac{dx^\mu}{d\tau}\frac{dx^\nu}{d\tau} = 0 .$$

Now multiply both sides of this by $\partial x^\lambda / \partial \xi^\alpha$ and sum over α:

$$\overbrace{\frac{\partial x^\lambda}{\partial \xi^\alpha}\frac{\partial \xi^\alpha}{\partial x^\nu}}^{\delta^\lambda_\nu}\frac{d^2 x^\nu}{d\tau^2} + \frac{\partial x^\lambda}{\partial \xi^\alpha}\frac{\partial^2 \xi^\alpha}{\partial x^\nu \partial x^\mu}\frac{dx^\mu}{d\tau}\frac{dx^\nu}{d\tau} = 0 .$$

Note that the product under the overbrace is 1 if $\lambda = \nu$ and 0 otherwise, which has the effect of turning d^2x^ν into d^2x^λ; so that

$$\frac{d^2 x^\lambda}{d\tau^2} + \frac{\partial x^\lambda}{\partial \xi^\alpha}\frac{\partial^2 \xi^\alpha}{\partial x^\nu \partial x^\mu}\frac{dx^\mu}{d\tau}\frac{dx^\nu}{d\tau} = 0 . \tag{5.5}$$

The first term in this expression is just the ordinary acceleration in coordinate x^λ and would be zero in an inertial reference frame; the complicated second term (a triple sum entailing 64 terms for any choice of λ) details the effects of gravitation on the particle's trajectory as expressed in an arbitrary coordinate system.

The effects of changing coordinate systems are entirely expressed in the terms containing ξ, so it is convenient to combine them by defining the **Affine Connection** Γ as[1]

$$\Gamma^\lambda_{\mu\nu} \equiv \frac{\partial x^\lambda}{\partial \xi^\alpha}\frac{\partial^2 \xi^\alpha}{\partial x^\nu \partial x^\mu} . \tag{5.6}$$

The Equations of Motion of a freely falling particle in an arbitrary coordinate system are then given as

$$\frac{d^2 x^\lambda}{d\tau^2} = -\Gamma^\lambda_{\mu\nu}\frac{dx^\mu}{d\tau}\frac{dx^\nu}{d\tau} . \tag{5.7}$$

In Appendix (A), Section A.1 we show that Γ may be expressed directly in terms of the metric tensor components as

[1] Sometimes called (mostly in older texts) the Christoffel Symbol or 3-Index Symbol. The name 'Affine Connection' is more descriptive of the mathematical nature of this creature, and thus will be used exclusively here. See Section A.1 for further discussion of this fundamental aspect of differential geometry.

$$\Gamma^{\lambda}_{\mu\nu} = \frac{1}{2} g^{\lambda\kappa} \left[\frac{\partial g_{\mu\kappa}}{\partial x^{\nu}} + \frac{\partial g_{\nu\kappa}}{\partial x^{\mu}} - \frac{\partial g_{\mu\nu}}{\partial x^{\kappa}} \right] . \tag{5.8}$$

That the affine connection incorporates metric tensor components is a consequence of the role played by curvature in the GR formulation of gravitation (to be explained further in the following chapter). Note that, in Cartesian coordinates in an inertial reference frame, all the metric tensor components are constants and thus $\Gamma^{\lambda}_{\mu\nu} = 0$ for all indices, implying zero acceleration. In this sense, the affine connection describes departures from inertial reference frames.

In summary: the Gravitational Equations of Motion in an arbitrary coordinate system $\{x\}$ are

Gravitational Equations of Motion

$$\frac{d^2 x^{\lambda}}{d\tau^2} = -\Gamma^{\lambda}_{\mu\nu} \frac{dx^{\mu}}{d\tau} \frac{dx^{\nu}}{d\tau} , \qquad \text{where} \tag{5.9}$$

$$\Gamma^{\lambda}_{\mu\nu} = \frac{1}{2} g^{\lambda\kappa} \left[\frac{\partial g_{\mu\kappa}}{\partial x^{\nu}} + \frac{\partial g_{\nu\kappa}}{\partial x^{\mu}} - \frac{\partial g_{\mu\nu}}{\partial x^{\kappa}} \right] , \tag{5.10}$$

and the $g_{\mu\nu}$ are the components of the metric tensor for the $\{x\}$ system. In applications to gravitation, where **g** incorporates the resulting space-time curvature, these two equations constitute the generally relativistic form of Newton's second law for conservative forces, $d^2\vec{x}/d\tau^2 = -\vec{\nabla}\Phi$, where Φ is the potential. We thus expect the components of the gravitational metric tensor to serve a role similar to that of gravitational potentials in Newtonian mechanics.

The form of the path thus described by a freely falling particle is not obvious from Equation (5.9). As shown in Section A.2, such trajectories are always geodesics (shortest distance between two points) in whatever geometry prevails; Equation (5.10) is commonly referred to as the **Geodesic Equation**. In non-Euclidean geometries (e.g., in the presence of gravitation) these will not be straight lines – on the surface of a sphere they are great circles, for example.

It is worth observing that the compact notation used here conceals a rather complex result. Written out in terms of the metric tensor and with explicit summations, the gravitational equations of motion are

$$\frac{d^2 x^{\lambda}}{d\tau^2} = -\frac{1}{2} \sum_{\mu} \sum_{\nu} \sum_{\kappa} g^{\lambda\kappa} \left[\frac{\partial g_{\mu\kappa}}{\partial x^{\nu}} + \frac{\partial g_{\nu\kappa}}{\partial x^{\mu}} - \frac{\partial g_{\mu\nu}}{\partial x^{\kappa}} \right] \frac{dx^{\mu}}{d\tau} \frac{dx^{\nu}}{d\tau} ,$$

which contains 64 terms, each of them containing three metric tensor derivatives.

5.1.2 The Newtonian Approximation

The acceleration of a freely falling object is given in an arbitrary coordinate system by Equation (5.9). This is a very general result in which the coordinates and metric tensor may derive from a gravitational field of any complexity, including a time-varying one. But the compactness of the tensor notation obscures the structure of the resulting motion: looking at Equation (5.9), what does the trajectory of a freely falling object look like? How could you tell? Some idea of how curved space-time produces particle trajectories can be gained by examining the special case wherein Newtonian gravitation is nearly correct. This will occur under the following circumstances.

1. All velocities are small compared to that of light.
2. The gravitational field is stationary (not changing with time).
3. The gravitational field is very weak.

The resulting equations of motion *must* reduce approximately to the Newtonian result,

$$\frac{d^2\vec{x}}{dt^2} \approx -\vec{\nabla}\Phi\,, \tag{5.11}$$

where Φ is the Newtonian gravitational potential. What this is saying is that the metric tensor $\mathbf{g}$ *must* have a form such that (5.11) will result whenever the gravitational field satisfies the above three classical (quasi-Newtonian) constraints; and since the metric tensor arises from space-time curvature this requirement sets useful constraints on the curvature that can arise from gravitating matter and energy.

In Appendix B we show that the quasi-Newtonian constraints on the Equations of Motion (5.9), (5.10), applied to the Newtonian Equation (5.11), require that the metric tensor satisfy

$$g_{00} \approx -\left(1 + \frac{2\Phi}{c^2}\right) \qquad \text{(weak, static field).} \tag{5.12}$$

This result is significant on many levels. First, it demonstrates the close relationship of metric tensor components to gravitational potentials. Second, it provides a quantitative measure of the departure of relativistic gravitation from Newtonian: relativistic effects may be expected to be important if the Newtonian gravitational potential Φ is *not* small compared with c^2. For a point mass in Newtonian gravity $\Phi = -GM/R$, so that

$$\begin{aligned} 2\Phi/c^2 &\approx -10^{-9} \quad \text{(surface of Earth)}\,, \\ &\approx -10^{-8} \quad \text{(surface of Sun)}\,, \\ &\approx -10^{-4} \quad \text{(surface of a typical white dwarf)}\,. \end{aligned}$$

Equation (5.12) provides a constraint on GR metrics that will be useful in derivation of the Einstein Field Equations in Chapter 7.

5.2 Strong Equivalence Principle

The WEP deals only with gravitational forces. The Strong Equivalence Principle (SEP), as enunciated by Einstein, includes all fundamental forces and claims that *by no means internal to a reference frame can one distinguish between acceleration of that frame, and the presence of a gravitational field.* Put succinctly,

> *All* of physics in a stationary reference frame in a gravitational field of acceleration g is equivalent to (exactly the same as) physics in a reference frame with no gravitational field, but accelerating at rate $-g$ relative to the stationary frame.

Another way of phrasing this principle is in terms of Locally Inertial Reference Frames:

> In any sufficiently small region of space-time it is possible to define a **Locally Inertial Reference Frame** (LIRF), in which there is no apparent curvature of space-time and the laws of physics are those of special relativity; i.e., in which there are no apparent gravitational affects.

From a geometrical point of view an LIRF is the equivalent of a locally flat space-time, and the above statement of the Strong Principle of Equivalence is evocative of one characteristic of the Riemannian geometries that underlie GR: any sufficiently small region in such a geometry must be locally flat, much as a square meter of ocean surface is indistinguishable from a flat plane (if the water is calm), even though we know it is a part of the surface of a sphere of very large radius. It is thus not surprising that Einstein chose Riemannian geometries for modelling the curvature of space-time.

An immediate significance of LIRFs for general relativity is that any *tensor* equation derived for an LIRF holds in *any* reference frame, due to the principle of general covariance. This is a very handy way to derive useful things: set up an LIRF at the point in space-time, derive a tensor equation of interest in the resulting inertial reference frame, and then feel free to use it in non-LIRF frames. This seems like cheating but it's not. It's not easy, either: the equation must be a tensor one, and not all things of interest can be so expressed; and the derived expression applies only locally.

This approach – which illustrates the power of General Covariance in combination with the SEP – will be used in Chapter 8 to model the overall expansion of the Universe in terms of SR physics. In addition to its contribution to GR *per se*, the SEP implies several connections between acceleration and gravitation that are of considerable interest.

5.2.1 Light deflection

Imagine two space ships – one at rest on a planet's surface, one accelerating in space – and send a light ray through each of them, perpendicular to the long axis of the ships. In the accelerating ship the light ray will appear to bend downward as it transits (in a finite time) the width of the accelerating ship; by the SEP the same bending must be observed in the stationary ship resting in a gravitational field. Thus: **gravitation deflects light** toward the gravitational source. This was the first prediction of GR to be experimentally verified, in observations of apparent star positions during the total solar eclipse of 1917. The paths of light rays from stars whose apparent positions were near the Sun were bent toward the Sun, resulting in the apparent shifting of the stellar positions away from the Sun. This was such a counterintuitive prediction of GR that its observation was a totally convincing verification of the then-new General Theory of Relativity, and catapulted Einstein into the public eye.

5.2.2 Gravitational redshift

Now consider the same two ships, this time with light rays along the axis from the back to the front of each ship. If the ship is of length L the light ray will take time $t = L/c$ to travel the length of the ship.[2] In the moving ship the velocity of the front end will have increased an amount $\Delta v = gt = gL/c$ during the light's travel from back to front, and the light ray will thus be received having been redshifted by amount $\Delta\lambda/\lambda = \Delta v/c = gL/c^2$. According to the SEP, the same result must hold in the stationary ship so that **gravitational fields redshift light.** The sense of the effect is to redshift light moving from lower to higher gravitational potentials: since the gravitational potential at the top of the ship is higher than that at the bottom by $\Delta\Phi = gL$, the gravitational redshift is given by

$$\frac{\lambda_o - \lambda_e}{\lambda_e} = \frac{\Delta\Phi}{c^2} \quad \left(= \frac{gL}{c^2}\right), \tag{5.13}$$

where λ_o is observed at a gravitational potential $\Delta\Phi$ above the level at which it was emitted with wavelength λ_e. An easy way to remember this is that light

[2] Since the moving ship is accelerating upward the light pulse travels a bit further than the length L, and there is a second-order correction to t of magnitude $L^2g^2/4c^2$, which will be *very* small unless either g or L are unreasonably large.

is redshifted as it climbs *out* of a gravitational potential well (and loses energy thereby).

The effect is observable, but not easily so. On the surface of the Sun the gravitational potential is lower than that far away from the Sun by approximately $\Phi/c^2 \approx 10^{-8}$, for a wavelength shift of less than 10^{-5} nm in visible light, equivalent to a radial velocity of a few meters per second – comparable to the precision (barely) achievable with modern radial velocity spectrographs that search for exoplanets. It has nonetheless been unambiguously observed in normal stars, white dwarfs (where the effect is larger by four orders of magnitude), and in the laboratory (a very tall one!).

5.2.3 Time dilation

The gravitational redshift is a consequence of *gravitational time dilation*. Imagine now that we send not a continuous light wave, but very short light pulses, from the stern to the bow of both ships. Since the speed of light is constant in both ships, the rate f at which the light pulses are received in the bow of the accelerating ship will be less than that at which they were emitted at the stern by $\Delta f/f = -\Delta v/c = -gL/c^2$. This implies that over time the total number of pulses received at the bow of the ship will be systematically less than the number emitted at the stern, a nonsensical result. Einstein reasoned that the solution to this apparent paradox was that the clocks at the top of the ship ran faster than those at the bottom, so that the *apparent* pulse rate was the same for both. By the SEP, the same must hold for the stationary ship in a gravitational field: clocks run more slowly at lower gravitational potentials. This is **gravitational time dilation**, in which time intervals vary with gravitational potential as

$$\Delta t_1 = \left(1 + \frac{\Phi_1 - \Phi_2}{c^2}\right) \Delta t_2 \,. \tag{5.14}$$

This, too, has been experimentally verified. In fact, correction for this effect is absolutely required in order to realize the precision of the Global Positioning System (GPS), in which the transmitting satellites are at a higher gravitational potential (Earth orbit) than are the Earth-bound receivers. Problem 3 at the end of this chapter invites you to verify this claim. It is a curious fact that this correction is probably the only effect associated with GR that enters (some of) our everyday lives.

Finally, note that the gravitational time dilation can equally be viewed as dilation due to acceleration of the reference frame; i.e., as **general relativistic time dilation**. Herein lies the real explanation of the twin paradox, where one twin flies off to Alpha Centauri at nearly the speed of light and returns home to find his/her twin sibling aged a comparatively large amount. The apparent paradox arises from the relativity of SR: since each twin has the same non-zero velocity with respect to the other, it seems as though they must age the same. But the flying twin is also accelerating during the flight, and the resulting additional time dilation accounts for the lesser aging.

Problems

1. If **g** is a diagonal metric tensor show that:

   (a) $\Gamma^i_{jk} = 0$ if $i \neq j \neq k \neq i$;
   (b) $\Gamma^i_{jj} = -\frac{1}{2} g^{ii} \left(\partial g_{jj} / \partial x^i \right)$ if $i \neq j$;
   (c) $\Gamma^i_{ij} = \frac{1}{2} g^{ii} \left(\partial g_{ii} / \partial x^j \right)$ for all i, j (no sum on i).

2. Use the results of Problem 1 to find all non-zero components of the affine connection for polar coordinates in flat space-time where the metric can be read from the invariant space-time interval:

$$ds^2 = -c^2 dt^2 + dr^2 + r^2 d\theta^2 + r^2 \sin^2 \theta \, d\phi^2 \, .$$

3. Global Positioning System (GPS) satellites orbit the Earth at a height where their orbital period is 12 hours, and are thus at a higher gravitational potential than the surface of the Earth. Assuming that the mean radius of the Earth is $R_\oplus = 6371$ km, and its mass is $M_\oplus = 5.97 \times 10^{24}$ kg, compute the potential difference between the Earth's surface and the satellites' orbits, and thus the GPS gravitational time dilation factor $(dt_1 - dt_2) / dt_2$. Use this to estimate the distance error in a GPS signal accumulated over several minutes, if the gravitational time dilation factor were not taken into account. (Since GPS communications are by radio the signals travel at the speed of light.) Comment on the necessity to correct GPS measurements for this time dilation.
4. A common metric employed on the two-dimensional surface of a 3-sphere of radius R is

$$ds^2 = R^2 d\theta^2 + R^2 \sin^2(\theta) \, d\phi^2,$$

   where θ is the colatitude and ϕ is the longitude.

   (a) Compute the non-zero components of the affine connection for this metric.
   (b) Use these components to show that the path of a freely falling object as seen in this reference frame is a great circle. Hint: use Equation (5.9) with a carefully chosen initial position and velocity.

5. Consider the usual fixed Cartesian coordinate system (x, y, z) with origin located on the Earth's surface and the z-axis pointing straight up. Close

to the surface the equation of motion of a falling object is $z(t) = z_0 - \frac{1}{2}gt^2$, where g is the usual gravitational acceleration, assumed constant. In terms of this fixed system, define a freely falling coordinate system $(\bar{x}, \bar{y}, \bar{z})$ centered on the falling object and show directly that $dz/d\tau = d^2\bar{z}/d\tau^2 = 0$ in this system.

6. Compute the relative time dilation factor between astronauts on the International Space Station orbiting the Earth at a height of $\sim$ 400 km, and stay-at-home types on the surface of the Earth. If an astronaut were to spend a year in orbit, by how much would he/she age relative to a twin that remains on Earth?

6

Space-time curvature

The conceptual basis of GR is that matter and energy cause space-time to be curved, and that curvature determines the paths of freely falling objects. From the previous chapter we expect that the curvature will be reflected in the affine connection or, more generally, in the metric tensor. It is thus necessary to define appropriate measures of curvature in terms of the metric tensor components.

Now, it is pretty easy to mathematically describe the curvature of, say, the two-dimensional surface of a sphere embedded in our three-dimensional space. But it's not at all obvious how to describe the curvature of the three-dimensional space in which we presumably live, nor even to understand what *curvature* means in that context. The trick is to use the concept of distance or, more precisely, the *metric* of the space. Then curvature will reveal itself by, e.g., the circumference of a circle in terms of its radius, the sum of the three angles in a triangle, etc. To be useful in application to GR we require that the curvature be revealed without moving outside the surface or space in question. Going back to our sphere embedded in 3-space, imagine a two-dimensional bug wandering around on the surface of the sphere, taking measurements. We need to develop the tools by which those measurements can be used to quantify the curvature of the surface, then generalize to four-dimensional space-time.

6.1 Simple curvature

Descriptions of curvature of plane curves and of two-dimensional surfaces are based on circles and spheres. The *curvature* K of a circle is defined to be the inverse of its radius R, which is called the *radius of curvature*: $K \equiv 1/R$. For any other plane curve the curvature at a point on the curve is defined to be that of the best-fitting circle to the curve at that point. In Cartesian coordinates, the curvature of a plane curve $y(x)$ at any chosen point is given by

$$K = \frac{1}{R} = \pm \left| \frac{y''}{\left(1 + y'^2\right)^{3/2}} \right| , \tag{6.1}$$

where the sign is chosen according to the chosen orientation. Note the presence of the second derivative, a characteristic of measures of curvature.

Extension of these ideas to a curved, two-dimensional surface in 3-space is fairly straightforward. At any point on a surface one can define principal curvatures R_1 and R_2 corresponding to curves resulting from intersections with tangent and normal planes and computed by Equation (6.1). The *Gaussian curvature* of the surface is then computed simply as

$$K_G = \pm \frac{1}{R_1} \frac{1}{R_2} , \tag{6.2}$$

where the sign is chosen according to the orientation of the surface: a convex surface is assigned a positive curvature, while a concave surface is assigned negative curvature. For uniformly curved surfaces, $R_1 = R_2 = R$ so that the Gaussian curvature of an external spherical surface is $K_G = 1/R^2$.

It can be shown – but not easily – that the Gaussian curvature of a two-dimensional surface can be computed by the Bertrand–Diquet–Puiseux formula

$$K_G = \lim_{r \to 0^+} 3 \frac{2\pi r - C(r)}{\pi r^3} , \tag{6.3}$$

where $C(r)$ is the circumference of a circle of radius r. It is apparent from this that the circumference of a circle of radius r is less than $2\pi r$ on the surface of a sphere ($K_G > 0$), and greater than that on a negatively curved surface. This formula is an elegant example of the use of measurements *on a surface* to determine its curvature, as opposed to measuring its radius as a figure embedded in a higher-dimensional space. We must do something similar to determine the curvature of our three-dimensional space from within it.

6.2 Uniformly curved surfaces

In analogy with the surface of a sphere in 3-space we can imagine three-dimensional space to be the surface of the four-dimensional analog. Since three- or four-dimensional surfaces embedded in higher dimensions are difficult to envision, we start with curvatures of two-dimensional surfaces and extend to higher dimensions by analogies. To keep the derivations at a reasonable level we will consider only surfaces of uniform curvature, by which we mean ones in which the curvature is the same at all places and in all directions – e.g., spheres, but not (American) footballs. We will generalize to spaces of arbitrary curvature at the end of the chapter.

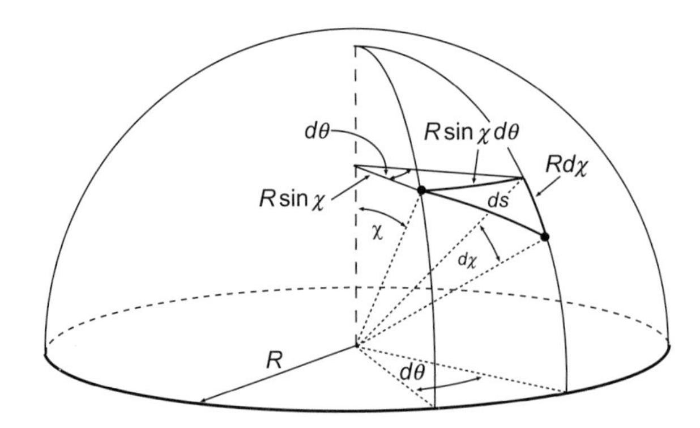

Figure 6.1 A coordinate system (χ, θ) on the two-dimensional surface of a sphere of radius R.

We start with curvature of two-dimensional surfaces. Figure 6.1 shows a simple coordinate system on a spherical surface, consisting of colatitude χ and longitude θ, for which the incremental distance ds between two closely spaced points is easily seen to be

$$ds^2 = R^2 d\chi^2 + R^2 \sin^2 \chi \, d\theta^2 \,. \tag{6.4}$$

Now change variables by $\rho \equiv R \sin \chi$, so $R d\chi = \pm d\rho / \sqrt{1 - (\rho/R)^2}$ and the above line element becomes

$$ds^2 = \frac{d\rho^2}{1 - (\rho/R)^2} + \rho^2 d\theta^2 \,. \tag{6.5}$$

To extend this to three dimensions we add the analog of a third polar coordinate, $\rho^2 \sin^2 \theta \, d\phi^2$; and to incorporate other forms of uniform curvature we generalize the first metric element according to

$$ds^2 = \frac{d\rho^2}{1 - K\rho^2} + \rho^2 d\theta^2 + \rho^2 \sin^2 \theta \, d\phi^2 \,, \tag{6.6}$$

where K is the **geometric curvature**: $K = +1/R^2$ for positively curved surfaces, $-1/R^2$ for negatively curved surfaces, and $K = 0$ for uncurved (flat) surfaces. The metric for this choice of coordinates is thus

$$\mathbf{g} = \begin{pmatrix} \left(1 - K\rho^2\right)^{-1} & 0 & 0 \\ 0 & \rho^2 & 0 \\ 0 & 0 & \rho^2 \sin^2 \theta \end{pmatrix} , \tag{6.7}$$

in terms of which the line element (6.6) may be written in its metric form $ds^2 = g_{ij} dx^i dx^j$ with $x^1 = \rho, x^2 = \theta, x^3 = \phi$.

That this generalization does indeed produce uniformly curved surfaces of positive, zero, and negative curvatures can be demonstrated by calculating the curvature from the Gaussian and Riemannian formulae presented later in this chapter. In positively curved geometries ρ is constrained to lie between 0 and R, while it can take on any value in flat and negatively curved spaces. This conforms to our visualization of three-dimensional objects: spheres ($K > 0$) have closed surfaces and saddles ($K < 0$) are open, and extend to infinity.

Note that other choices of coordinates will lead to other metrics, but will not affect the geometry (which is invariant with respect to changes in coordinate systems). A commonly employed alternative is based on Equation (6.4) with the radial coordinate defined as $r \equiv R\chi$, which leads to the metric

$$ds^2 = dr^2 + S_k\left(r\right)^2 \left[d\theta^2 + \sin^2\left(\theta\right)\, d\phi^2\right], \tag{6.8}$$

where $S_k\left(\mathfrak{r}\right) = R\sin\left(r/R\right)$ for positive curvature, $R\sinh\left(r/R\right)$ for negative curvature, and r for flat geometries. Either choice of coordinate system and metric will suffice to describe spaces of uniform curvature (as will many others), but will entail different intermediate results that can be confusing unless one selects a coordinate system and sticks with it – as we shall do in this text, with Equations (6.6) and (6.7) as the system of choice.

An implication of this result is that while the geometry of a surface may be embedded within the metric tensor, the embedding is not unique: different metric tensors (arising from different choices of coordinates) can represent the same geometry. This makes it difficult to extract the curvature from the metric tensor alone, which is a potential problem for GR where the densities of mass and energy will be used to infer not the curvature *per se*, but only the metric tensor via the Einstein Field Equations. What is now needed is some way to compute curvature from the metric tensor components themselves.

6.3 Metric measures of curvature

6.3.1 Gaussian curvature

In the eighteenth century Carl Gauss discovered the exact connection between metric tensor components and the curvature of two-dimensional surfaces. For coordinate systems with diagonal metric tensors Gauss' *Theorema Egregium* ('remarkable theorem') states that the Gaussian curvature (Equation (6.2)) of a surface is

$$K_G = \frac{1}{2g_{11}g_{22}} \left\{ -\frac{\partial^2 g_{11}}{\left(\partial x^2\right)^2} - \frac{\partial^2 g_{22}}{\left(\partial x^1\right)^2} + \frac{1}{2g_{11}} \left[\frac{\partial g_{11}}{\partial x^1} \frac{\partial g_{22}}{\partial x^1} + \left(\frac{\partial g_{11}}{\partial x^2} \right)^2 \right] + \frac{1}{2g_{22}} \left[\frac{\partial g_{11}}{\partial x^2} \frac{\partial g_{22}}{\partial x^2} + \left(\frac{\partial g_{22}}{\partial x^1} \right)^2 \right] \right\} . \tag{6.9}$$

This formula is not so much useful as it is illustrative. Note in particular the derivatives of the metric components, which essentially contain the shape information for the geometry: the expression is homogeneously quadratic in derivatives, so that each term is either the product of two first derivatives or is a second derivative. We expect to see the same sort of formulation when we employ space-time curvature in the Einstein Field Equations: non-linear combinations of the metric tensor components and their first and second derivatives. One can trivially show that this formula yields $K_G = 0$ for both the Cartesian and polar forms of the flat-geometry metric tensor; it's a bit more of a challenge (but not much) to show that $K_G = 1/R^2$ for the closed, spherical geometry on the surface of a sphere of radius R.

Curvature of plane curves is characterized by a single number, while curvature of two-dimensional surfaces requires two numbers (e.g., R_1 and R_2 of Equation (6.2); uniformity of curvature reduces this to one number). Curvature of four-dimensional surfaces, including that of space-time, requires 20 independent numbers to fully characterize it. The extension of Gauss' Theorema Egregium to more than two dimensions was provided by his student Riemann in the form of what is now called the Riemann Curvature Tensor.

6.3.2 Riemannian curvature

Riemann chose to characterize curvature in terms of measurable quantities on curved surfaces; familiar examples would be the relation between radii and circumferences of circles (e.g., Equation (6.3)) or the sum of interior angles in triangles. Perhaps the easiest way to envision curvature and its characterization is in terms of *parallel transport*, by which is meant movements of vectors along geodesics so that they always stay parallel to themselves.

Thus: consider moving a vector over the surface of a sphere in 3-space. Start at the north pole with a vector V pointing straight down the meridian and move the vector down to the equator without rotating it. Now move it along the equator by 90 degrees, again keeping it always parallel to itself (pointing straight south). Finally, move it back up its new meridian to the pole, and observe that it will have been rotated 90 degrees with respect to its original orientation, even though it was moved parallel to itself all along its path (i.e., its orientation after moving an infinitesimal distance was always parallel to its starting position). This rotation upon parallel transport (*holonomy*) occurs if, and only if, the surface is curved.

Riemann thus took as a measure of curvature of any space the amount of vector rotation ΔV upon parallel transport about an infinitesimal closed path, $\Delta f^{\mu\kappa}$, bounded by coordinates μ and κ. He showed that the change in vector components thus induced could be written as

$$\Delta V_\nu = \mathcal{R}^\lambda_{\ \mu\nu\kappa} V_\lambda \, \Delta f^{\mu\kappa}$$

(in the limit $\Delta f^{\mu\kappa} \to 0$); where $\mathcal{R}^\lambda_{\ \mu\nu\kappa}$ is the **Riemann Curvature Tensor** given by[1]

$$\mathcal{R}^\lambda_{\ \mu\nu\kappa} = \frac{\partial \Gamma^\lambda_{\mu\nu}}{\partial x^\kappa} - \frac{\partial \Gamma^\lambda_{\mu\kappa}}{\partial x^\nu} + \Gamma^\eta_{\mu\nu}\Gamma^\lambda_{\kappa\eta} - \Gamma^\eta_{\mu\kappa}\Gamma^\lambda_{\nu\eta} \,. \tag{6.10}$$

This tensor (it *is* a tensor, even if we haven't proven it) fully characterizes curvature: space-time is flat if, and only if, all components of the Riemann Tensor are zero in all coordinate systems. See Section A.4 for the derivation of this curvature tensor and a discussion of its properties.

For most applications this 256-component tensor is overkill and can be replaced by its rank-2 contraction, the **Ricci Tensor**:

$$\mathcal{R}_{\mu\kappa} \equiv \mathcal{R}^\lambda_{\ \mu\lambda\kappa} \tag{6.11}$$

$$= \frac{\partial \Gamma^\lambda_{\mu\lambda}}{\partial x^\kappa} - \frac{\partial \Gamma^\lambda_{\mu\kappa}}{\partial x^\lambda} + \Gamma^\eta_{\mu\lambda}\Gamma^\lambda_{\kappa\eta} - \Gamma^\eta_{\mu\kappa}\Gamma^\lambda_{\lambda\eta} \,. \tag{6.12}$$

This can be contracted further to the **curvature scalar**:

$$\mathcal{R} = g^{\mu\nu}\mathcal{R}_{\mu\nu} \,. \tag{6.13}$$

These two are the only useful contractions of the Riemann Tensor: all others are equivalent to them, or identically zero.

The compact tensor notation obscures the complexity of these curvature tensors. Each of the 16 components of the Ricci Tensor, for instance, is a double sum of Affine Connections and their derivatives, altogether entailing more than 4000 terms in the metric tensor components; see Equation (7.5) for an explicit listing.

[1] The sign of this tensor is arbitrary; the choice made here – and perpetuated into the Ricci tensor – is that of Weinberg (2008).

Problems

1. Using the usual cylindrical coordinate system, write down the metric tensor for the surface of a right circular cylinder of radius R. Then apply the Gaussian curvature formula (Equation (6.9)) to compute the Gaussian curvature of the surface. You might want to guess the answer first.
2. Using the usual spherical polar coordinates for the surface of a sphere of radius R, as given in Equation (2.1), use the Gaussian curvature formula to show that its Gaussian curvature is R^{-2}.
3. Compute the Gaussian curvatures of two-dimensional surfaces of constant curvature by applying Equation (6.9) to the two-dimensional metric equation

$$ds^2 = \frac{d\rho^2}{1 - K\rho^2} + \rho^2 d\theta^2$$

for the three cases $K > 0$, $= 0$, and < 0; and thereby verify that the metric Equation (6.6) does indeed represent all three types of uniform curvature of two-dimensional surfaces.
4. Compute the Gaussian curvature of a plane surface in both Cartesian and polar coordinates, and show that both coordinate systems give the same result (and that, therefore, different metrics can encode the same curvature).
5. Show that the formula for the circumference of a circle drawn on the surface of a sphere, as derived in Problem 1 of Chapter 2, conforms to the Bertrand–Diquet–Puiseux formula (Equation (6.3)).

7

Einstein Field Equations of gravitation

Space-time curvature is established by the density of gravitating matter and energy, and is reflected in the metric tensor components. The equations relating the resulting metric tensor to matter/energy density are the **Einstein Field Equations (EFE)** of gravitation; generically, $\mathcal{G}(\mathbf{g}) = T(\rho, \varepsilon)$ where T characterizes the density of sources of gravitation and $\mathcal{G}$ is a curvature tensor containing metric tensor components and their derivatives.

The Newtonian equivalent of the EFE is Poisson's Equation, $\nabla^2\Phi = 4\pi G\rho$, where Φ is the gravitational potential and ρ is matter density. This equation may be logically deduced from Newton's laws of motion and of gravitation, but Einstein's Field Equations are less secure: the relation between space-time curvature and mass/energy cannot be unambiguously deduced from fundamental principles, but must be inferred – guessed at, if you like – from broad principles and analogies that leave room open for many specific possibilities. It was arguably Einstein's greatest contribution that he came up with what appears to be the right answer: his field equations have stood the test of time (so far) and are the basis for relativistic cosmology (among other things).

Of the two halves of the EFE the curvature side is the more problematic, so we begin the discussion with the tensor representing matter and energy, which can – to a large extent – be logically deduced from familiar physics.

7.1 Sources of gravitation

7.1.1 Energy-momentum tensor

Possible sources of gravitation include *all* forms of mass/energy, including such things as kinetic energy, pressure (which carries dimensions of energy density), stress, electromagnetic field energies, etc.; in addition to 'normal' matter and radiation. One consequence is that the tensor describing the densities of sources of gravitation is known by many names: mass-energy tensor, energy-momentum tensor, stress-energy tensor, etc. We will stick to the most commonly used name, the **energy-momentum tensor**, and denote it by $\mathcal{T}$.

It is not hard to guess that this tensor must be of rank 2 if it is to include directional components such as momenta, stress, and electromagnetic field potentials. The proper and most general definition of $\mathcal{T}$ is a variational one: $\mathcal{T}$ is the quantity needed for a stationary matter/radiation action.[1] A more practical way to construct a plausible form for this tensor, when only mechanical forms of energy and momentum are included, is in terms of the special relativistic 4-momentum vector $\mathbf{P} = (E/c, \vec{p})$. In this formulation $\mathcal{T}^{\mu\nu}$ is the flux of $\mathbf{P}^{\mu}$ across a surface whose normal is in the νth direction, and thus:[2]

$$\begin{aligned}
\mathcal{T}^{00} &= \text{energy density}, \\
\mathcal{T}^{i0} &= c \times \text{density of } i\text{th component of momentum} \quad (i = 1,2,3)\,, \\
\mathcal{T}^{0j} &= c^{-1} \times \text{energy flux in the } j\text{th direction} \quad (j = 1,2,3)\,, \\
\mathcal{T}^{ij} &= \text{flux in } i\text{th direction of } j\text{th component of momentum} \\
&\qquad (i,j = 1,2,3).\ \text{It follows that} \\
\mathcal{T}^{ii} &= \text{pressure in the } i\text{th direction} \quad (i = 1,2,3)\,.
\end{aligned} \tag{7.1}$$

Note that all components of this tensor have dimensions of energy density, and that the pressure referred to in the last line is pressure *on* the system *by* its surroundings. This is, of course, just the manner in which pressure is normally interpreted in, e.g., the first law of thermodynamics: $dE = -PdV + \cdots$, so that work is done *on* the system in its compression. A system in equilibrium with its surroundings exerts that same pressure back, but in dynamic situations the pressure can be unbalanced. In cosmological applications the sources of pressure are radiation and (probably) dark energy; galaxies interact too weakly to meaningfully contribute to the pressure.

7.1.2 Conservation relations

In flat space-time where SR strictly holds, conservation of mass/energy and momentum is expressed as

$$(\mathrm{Div}T)^{\nu} = (\nabla T)^{\nu} = \frac{\partial T^{\mu\nu}}{\partial x^{\mu}} = 0\,;$$

[1] Explicit variational formulations for $\mathcal{T}$ can be found in, e.g., Section 4.3 of Carroll (2004), Appendix B.9 of Weinberg (2008), and Section 2.8 of Narlikar (1993).

[2] See, e.g., Section 7.5 of Collier (2013) or Section 22.1 of Hartle (2003) for derivation of these relations.

i.e., the flux of each component $\mathbf{P}^{\nu}$ of the momentum 4-vector is conserved: $\nu = 0$ for mass/energy, $\nu = 1, 2, 3$ for the three spatial components of momentum. Formally, this same relation holds in curved space-time where the derivative is covariant:

$$(\nabla T)^{\nu} = D_{\mu} T^{\mu\nu} = 0 \,. \tag{7.2}$$

In common parlance, the energy-momentum tensor is said to be *divergence-free*. But in curved space-time this does not imply mass/energy conservation. One way to understand this is to note that $T^{\mu\nu}$ does not include gravitational energy *per se*: gravitating mass/energy creates a gravitational field which represents an energy component in the form of space-time curvature (the equivalent of gravitational potential energy in SR), that is not included in $T^{\mu\nu}$. Partially as a consequence, the concept of energy conservation is poorly defined in curved space-time, which is a dynamical partner to matter and energy (as opposed to the situation in SR where space-time is a non-interacting background).

7.2 Field Equations

7.2.1 The Einstein Tensor

If the energy-momentum tensor is to form one side of the equality constituting the field equations, the other side must be a second-rank, divergence-free tensor containing curvature; and it must be of such a form that the EFE reduce to Newtonian gravitation in the static, weak-field limit. The simplest choices for the curvature tensor are the metric tensor itself, **g**, and the Ricci tensor, $\mathcal{R}_{..}$. But the Ricci tensor is not divergence-free, and the metric tensor yields a form for the EFE that does not reduce to Newtonian. So Einstein chose for the curvature side of the field equations the simplest possible combination of these two that satisfies both requirements:

$$\mathcal{G}_{\mu\nu} \equiv \mathcal{R}_{\mu\nu} - \frac{1}{2} g_{\mu\nu} \mathcal{R} \,, \tag{7.3}$$

where $\mathcal{R} \equiv g^{\lambda\kappa} \mathcal{R}_{\lambda\kappa}$ is the curvature scalar (Equation (6.13)). This is the **Einstein Tensor**; that it is divergence-free is shown in Section A.6.

7.2.2 Einstein Field Equations (EFE)

Einstein initially chose for his field equations the simplest way in which the Einstein Tensor and the energy-momentum tensor could be combined so as to reduce to Newtonian gravitation in the static, weak-field limit:

$$\mathcal{G}_{\mu\nu} \equiv \mathcal{R}_{\mu\nu} - \frac{1}{2} g_{\mu\nu}\mathcal{R} = -\frac{8\pi G}{c^4}\mathcal{T}_{\mu\nu}\,, \tag{7.4}$$

where $\mathcal{T}_{\mu\nu} = g_{\mu\kappa}g_{\lambda\nu}\mathcal{T}^{\kappa\lambda}$ is the covariant form of the energy-momentum tensor, and the constant multiplying $\mathcal{T}_{\mu\nu}$ is needed for compliance in the Newtonian limit (as shown in Appendix B). These constitute a set of second-order differential equations in the metric tensor components, the solutions to which determine gravitational dynamics of freely falling objects via such equations as (5.9).

The field equations are 16 linked, partial differential equations in the 16 metric tensor components.[3] Each such equation is a very complicated combination of the $g_{\kappa\lambda}$ and their derivatives. The tensor notation allows us to write such things in a very compact manner, but it is worth keeping in mind how complex these curvature tensors really are. The Ricci Tensor, written directly in terms of the $g_{\mu\nu}$ (eschewing the notational convenience of the affine connection), and with summations made explicit, is

$$\begin{aligned}
\mathcal{R}_{\mu\nu} = &\frac{1}{2}\sum_{\lambda}\sum_{\kappa}\left[\frac{\partial g^{\kappa\lambda}}{\partial x^{\nu}}\left(\frac{\partial g_{\mu\kappa}}{\partial x^{\lambda}} + \frac{\partial g_{\lambda\kappa}}{\partial x^{\mu}} - \frac{\partial g_{\mu\nu}}{\partial x^{\kappa}}\right)\right.\\
&\left.+g^{\kappa\lambda}\left(\frac{\partial^2 g_{\mu\kappa}}{\partial x^{\lambda}\partial x^{\nu}} + \frac{\partial^2 g_{\lambda\kappa}}{\partial x^{\mu}\partial x^{\nu}} - \frac{\partial^2 g_{\mu\lambda}}{\partial x^{\nu}\partial x^{\kappa}}\right)\right]\\
&-\frac{1}{2}\sum_{\lambda}\sum_{\kappa}\left[\frac{\partial g^{\kappa\lambda}}{\partial x^{\lambda}}\left(\frac{\partial g_{\mu\kappa}}{\partial x^{\nu}} + \frac{\partial g_{\nu\kappa}}{\partial x^{\mu}} - \frac{\partial g_{\mu\nu}}{\partial x^{\kappa}}\right)\right.\\
&\left.+g^{\kappa\lambda}\left(\frac{\partial^2 g_{\mu\kappa}}{\partial x^{\nu}\partial x^{\lambda}} + \frac{\partial^2 g_{\nu\kappa}}{\partial x^{\mu}\partial x^{\lambda}} - \frac{\partial g_{\mu\nu}}{\partial x^{\kappa}\partial x^{\lambda}}\right)\right]\\
&+\frac{1}{4}\sum_{\lambda}\sum_{\eta}\left\{\left[\sum_{\kappa} g^{\eta\kappa}\left(\frac{\partial g_{\mu\kappa}}{\partial x^{\lambda}} + \frac{\partial g_{\lambda\kappa}}{\partial x^{\mu}} - \frac{\partial g_{\mu\lambda}}{\partial x^{\kappa}}\right)\right]\right.\\
&\left.\times\left[\sum_{\xi} g^{\lambda\xi}\left(\frac{\partial g_{\nu\xi}}{\partial x^{\eta}} + \frac{\partial g_{\eta\xi}}{\partial x^{\nu}} - \frac{\partial g_{\nu\eta}}{\partial x^{\xi}}\right)\right]\right\}\\
&-\frac{1}{4}\sum_{\lambda}\sum_{\eta}\left\{\left[\sum_{\kappa} g^{\eta\kappa}\left(\frac{\partial g_{\mu\kappa}}{\partial x^{\nu}} + \frac{\partial g_{\nu\kappa}}{\partial x^{\mu}} - \frac{\partial g_{\mu\nu}}{\partial x^{\kappa}}\right)\right]\right.\\
&\left.\times\left[\sum_{\xi} g^{\lambda\xi}\left(\frac{\partial g_{\lambda\xi}}{\partial x^{\eta}} + \frac{\partial g_{\eta\xi}}{\partial x^{\lambda}} - \frac{\partial g_{\lambda\eta}}{\partial x^{\xi}}\right)\right]\right\}.
\end{aligned} \tag{7.5}$$

[3] Only 10 of which are independent due to symmetry of the Ricci Tensor in its indices.

The first two lines alone imply $2 \times 3 \times 4 \times 4 = 96$ individual products of metric tensors and derivatives when multiplied out through the double sum, and the last two lines – involving triple sums – have 2, 304 terms each: altogether evaluation of the Ricci Tensor for any one choice of indices (μ, ν) requires computation of 4, 800 terms, each of them either the product of two first derivatives of metric tensor components, or a second derivative.

The curvature scalar $\mathcal{R}$, involving another double sum, is an order of magnitude more complex, so it is worthwhile to manipulate the field equations in order to eliminate it. Collapsing indices throughout Equation (7.4) with $g^{\mu\nu}$,

$$\mathcal{R} - \frac{1}{2}\, 4\mathcal{R} = -\,\frac{8\pi G}{c^4}\mathcal{T}\,,$$

where $\mathcal{T} = g^{\mu\nu}\mathcal{T}_{\mu\nu}$ is the energy-momentum scalar. Then $\mathcal{R} = \left(8\pi G/c^4\right)\mathcal{T}$ which, substituted back into the field equations, yields the alternative form

$$\mathcal{R}_{\mu\nu} = -\frac{8\pi G}{c^4}\left(\mathcal{T}_{\mu\nu} - \frac{1}{2}g_{\mu\nu}\mathcal{T}\right). \tag{7.6}$$

This is the most commonly used version of the basic EFE.

7.2.3 Cosmological Constant (I)

Einstein's initial applications of the Field Equations were to the deflection of starlight by the Sun and the precession of Mercury's orbit. In both of these his equations were successful in predicting observable gravitational effects that were inexplicable with Newtonian gravitation, and helped to solidify this theory in the eyes of the scientific community. But his initial attempts to model the dynamics of the entire Universe failed, and led him to (temporarily) modify the field equations to account for cosmological dynamics.

At the time ($\sim$ 1915) the Universe was thought to be filled by stars that were, on the whole, motionless with respect to the Sun: external galaxies were as yet unrecognized as such. But gravitation is an entirely attractive force so a static assembly of gravitating objects was not possible with either Newtonian or relativistic gravitation: the Universe would have to be either expanding or contracting. Since it was then thought to be doing neither, Einstein was compelled to add a term to the Field Equations that corresponded to a force of repulsion that would balance gravitation.

In one sense this turned out to be an easy modification:

$$\mathcal{R}_{\mu\nu} - \frac{1}{2}g_{\mu\nu}\mathcal{R}\; \overbrace{\underbrace{-g_{\mu\nu}\Lambda}} = -\frac{8\pi G}{c^4}\mathcal{T}_{\mu\nu}\,, \tag{7.7}$$

where Λ is a constant.[4] This modified form of the field equations is mathematically correct – both sides are divergence-free – but it does not reduce to Newtonian gravitation in the weak-field limit, unless Λ is very small. Since the modification is needed only on cosmological scales a very small value of Λ will suffice, so Einstein called Λ the **Cosmological Constant** – as it is still known today. It has no appreciable effect in application on non-cosmological scales, such as with black holes.

When Hubble and others discovered the expansion of the Universe in the 1920s Einstein gave up on this inelegant addition to his basic Field Equations, which he then characterized as being too ugly to be realized in nature. The constant was re-introduced in the late 1990s when observations of distant galaxies suggested that the Universe's expansion was accelerating, something allowed by a cosmological constant of proper value (among other possibilities). Its inclusion in the Field Equations – Equation (7.7) – has since become standard practice because, among other things, models derived with its inclusion can always be reduced to those of the original field equations simply by setting Λ to zero. The reduced form of the EFE including Λ – replacing Equation (7.6) – is

$$\mathcal{R}_{\mu\nu} = -\frac{8\pi G}{c^4}\left(\mathcal{T}_{\mu\nu} - \frac{1}{2}g_{\mu\nu}\mathcal{T}\right) - g_{\mu\nu}\Lambda \ . \tag{7.8}$$

This form is the most commonly used on cosmological scales.

7.3 Summary

At this point we have derived the following results: first, that the dynamical effect of gravitation on an object is to cause it to follow a geodesic path in curved spacetime given by

$$\frac{d^2x^\lambda}{d\tau^2} + \Gamma^\lambda_{\mu\nu}\frac{dx^\mu}{d\tau}\frac{dx^\nu}{d\tau} = 0$$

(Equation (5.9)), where

$$\Gamma^\sigma_{\mu\lambda} = \frac{1}{2}g^{\nu\sigma}\left[\frac{\partial g_{\mu\nu}}{\partial x^\lambda} + \frac{\partial g_{\lambda\nu}}{\partial x^\mu} - \frac{\partial g_{\mu\lambda}}{\partial x^\nu}\right]$$

(Equation (5.10)); i.e., motion is determined by the metric tensor components and their derivatives. Second, these components must conform to the distribution of

[4] That this addition corresponds to a force of repulsion will be shown in Section 8.4.1.

gravitating mass/energy, $\mathcal{T}_{\mu\nu}$, according to the Einstein Field Equations,

$$\mathcal{R}_{\mu\nu} = -\frac{8\pi G}{c^4}\left(\mathcal{T}_{\mu\nu} - \frac{1}{2}g_{\mu\nu}\mathcal{T}\right) - g_{\mu\nu}\Lambda\ , \tag{7.9}$$

(Equation (7.8)), where the Ricci Tensor is a complicated function of the metric tensor components and their first two derivatives:

$$\mathcal{R}_{\mu\kappa} = \frac{\partial\Gamma^{\lambda}_{\mu\lambda}}{\partial x^{\kappa}} - \frac{\partial\Gamma^{\lambda}_{\mu\kappa}}{\partial x^{\lambda}} + \Gamma^{\eta}_{\mu\lambda}\Gamma^{\lambda}_{\kappa\eta} - \Gamma^{\eta}_{\mu\kappa}\Gamma^{\lambda}_{\lambda\eta} \tag{7.10}$$

(Equation (6.12)). Given the distribution $\mathcal{T}_{\mu\nu}$ of mass and energy densities, and an estimate for the cosmological constant Λ, these equations constitute a well-posed problem that can, in principle, be solved for the motion of a particle in a gravitational field of any sort.

The practical difficulty is that this set of equations is far too complex to solve by ordinary means. The Field Equations themselves constitute ten linked, second-order, non-linear partial differential equations in the ten independent metric tensor components, each of them entailing several thousand terms for each choice of indices on $\mathcal{R}_{\mu\kappa}$. Fortunately, in most simple astronomical applications many of the metric tensor components are zero or redundant, but the whole set of equations is still dauntingly large and complex. It is probably possible – one suspects that NASA or ESA might already have done it – to program the Field Equations in all their generality for numerical solutions given, say, whatever mass/energy density distribution one chooses, and initial position and velocity of the object of interest. But this sort of brute force approach is not very satisfactory if one is interested in the larger picture of how, e.g., the trajectory of an object or light ray near a gravitating object varies from the Newtonian prediction. What the typical astronomer needs is a way of inferring analytical solutions (if only approximate ones) to the Field Equations for the simplest mass/energy density distributions encountered in practice: a black hole, a homogeneous and isotropic Universe, a spherically symmetric star or galaxy cluster, etc.

In this book we will concern ourselves only with cosmological applications of the Field Equations, and even there we will consider only the simplest and most symmetric possibility: a perfectly uniform and isotropic distribution of gravitating mass/energy. Extensions to non-uniform universes are the subject of a great deal of current study, but are at least an order of magnitude more difficult to model and so will not be dealt with here (aside from a first-order analysis of small irregularities in Chapters 17 and 18).

Part III

Universal expansion

For those skipping Part II

Here is a synopsis of the GR background to the expansion equations of the Universe.

The dynamics of universal expansion are described by the Einstein Field Equations, the relativistic equivalent of Poisson's Equation of Newtonian gravitation, $\nabla^2\Phi = 4\pi G\rho$. In GR, the equivalent of gravitational potential Φ is spacetime curvature as described by the curvature tensor $\mathcal{R}$, and the equivalent of mass density ρ is the energy-momentum tensor $\mathcal{T}$. The GR field equations then become

$$\mathcal{R}_{\mu\nu} = -\frac{8\pi G}{c^4}\left(\mathcal{T}_{\mu\nu} - \frac{1}{2}g_{\mu\nu}\mathcal{T}\right) - g_{\mu\nu}\Lambda \,, \tag{III.1}$$

where $g_{\mu,\nu}$ is the metric tensor describing the coordinate system in use and Λ is the cosmological constant, a 'fudge factor' introduced by Einstein to conform his equations to the Universe as it then appeared to him. The curvature tensor $\mathcal{R}_{\mu\nu}$, known as the Ricci Tensor, is a complicated function of the metric tensor components $g_{\mu\nu}$ and their first two derivatives.

The metric tensor components serve as GR generalizations to the SR spacetime invariant describing distances measured in inertial reference frames:

$$ds^2 = g_{00}\left(x^0\right)^2 + g_{11}\left(x^1\right)^2 + g_{22}\left(x^2\right)^2 + g_{33}\left(x^3\right)^2 \,, \tag{III.2}$$

where x^0 plays the role of time and $\left(x^1, x^2, x^3\right)$ are the spatial coordinates. The Einstein Field Equations (III.1) consist of 16 linked, second-order, non-linear differential equations in the metric tensor components $g_{\mu\nu}$ that must be solved for those components, given the distribution of gravitating matter and energy described by $\mathcal{T}$ and Λ. These metric tensor components are then used in the cosmological form of Equation (III.2) to describe gravitational dynamics, much as Newtonian gravitational potentials are used in Newtonian dynamics.

The following chapter sets up the appropriate metric for cosmology and then solves for the metric tensor components as expressed for an expanding Universe. Of central importance is the employment of a coordinate system that fully exploits the symmetries of an expanding Universe and that reduces the very complicated Einstein Field Equations to relatively simple equations that can be applied to the Universe we observe. The remaining chapters in this section then connect the cosmological metric to observable properties of the Universe.

8

Cosmological Field Equations

8.1 Cosmological coordinates

Since the Field Equations are generally covariant we are free to choose the coordinate system in which to express them in the manner that best suits our purposes. We do so here in a form that fully exploits the symmetries of cosmology, and that represents the expanding Universe as an inertial reference frame in which the effects of gravitation are subsumed within the coordinate system itself. This requires invocation of two principles: the Cosmological Principle and a form of the Equivalence Principle.

8.1.1 The Cosmological Principle

The **Cosmological Principle** asserts that the Universe is homogeneous and isotropic, the same in all places and in all directions. This principle is well supported by observations on sufficiently large scales, those of $\sim$ 150 Mpc or greater. On smaller scales it is clearly violated by gravitational clumping of galaxies into clusters and superclusters, but for the Universe as a whole it seems to be a valid extension of the Copernican Principle, that there is nothing special about our particular place in the scheme of things. The Cosmological Principle is a strong claim of symmetry and implies, among other things, that the curvature of space-time is uniform throughout the Universe at any given time (although it may change with time). From Chapter 6 the spatial portion of the Universe's metric is thus describable by the simple line element for three-dimensional spaces of constant curvature:

$$d\sigma^2 = \frac{d\rho^2}{1 - K\rho^2} + \rho^2 d\theta^2 + \rho^2 \sin^2\theta \, d\phi^2 \, , \tag{8.1}$$

where $K = \left(-1/\left|R\right|^2,\ 0,\ +1/\left|R\right|^2\right)$ if the curvature is (negative, zero, positive), respectively; with R playing the role of radius of curvature. This describes a diagonal metric and so is much simpler than the full 3×3 array needed for the most general form. The issue now becomes one of incorporating time into the metric.

8.1.2 Co-moving coordinates

You can envision the expanding Universe as a system in which the galaxies are freely falling (upward) in a gravitational field. The situation is not unlike that of Einstein's workmen falling from the building roof: on the way down their local reference frame is essentially inertial since everything in it is falling with the same acceleration. Similarly, we can transform the expanding Universe into an inertial reference frame by adopting a freely falling coordinate system in which *the spatial coordinates of a galaxy do not change as a consequence of expansion.* In the commonly used parlance this is a **co-moving coordinate system**; the homey analogy is to that of raisins embedded in a rising loaf of bread dough, or ink dots on an inflating balloon: we can assign an unchanging coordinate to each 'galaxy' and keep track of changing distances, etc., entirely in terms of expansion of the underlying coordinate grid.

We can picture co-moving coordinates in terms of galaxy world lines, as in SR and as shown in Figure 8.1. The spatial coordinates $x^{\{1,2,3\}}$ of any galaxy (solid curve) do not change along its world line, while the geometry of the bundle of world lines corresponding to a set of galaxies evolves with time. **Time** in such a system is defined in terms of hypersurfaces connecting the bundle of world lines (dashed lines marked with time coordinate x^0), and is commonly called **cosmic** or **coordinate** time.

The angular coordinates (θ, ϕ) in the spatial metric (8.1) do not change with uniform expansion of an isotropic medium, and so are automatically co-moving. But the radial variable ρ is a function of time that monotonically increases with

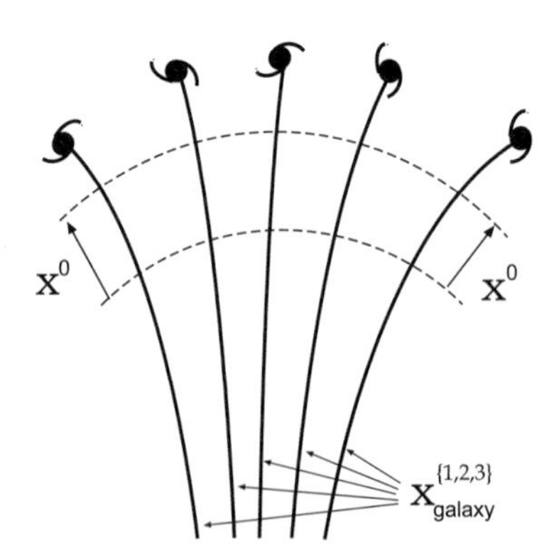

Figure 8.1 World lines of galaxies in a co-moving coordinate system corresponding to an expanding Universe: the spatial coordinates x^i do not change along any world line. The time coordinates are defined as hypersurfaces (dashed line) connecting the entire bundle of galaxy world lines.

universal expansion, so we replace it with the co-moving $r \equiv \rho(t)/a(t)$ where $a(t)$ is a universal expansion function as introduced in the Introduction (see Equation (I.1) ff) and normalized to $a(t_0) = 1$. Since $K \propto R^{-2}$ and $R \propto a$, the metric curvature term $K\rho^2$ becomes the invariant $K_0 r^2$, where K_0 is the current value of the curvature, and the co-moving metric is thus

$$ds^2 = g_{00}(cdt)^2 + a(t)^2 \left[\frac{dr^2}{1 - K_0 r^2} + r^2 d\theta^2 + r^2 \sin^2\theta \, d\phi^2 \right], \tag{8.2}$$

with ct taken as x^0 in conformance with SR.

Think of this coordinate system as describing an expanding grid, infinite in extent, whose interstices are labelled with their co-moving coordinates (r, θ, ϕ) that do not change as the grid expands. In such a system the coordinate origin is entirely arbitrary, as are the orientations of the θ and ϕ axes. The galaxies, also being co-moving, do not change their coordinates as the Universe expands. The consequences of this choice of coordinates for descriptions of cosmological dynamics are profound.

8.1.3 Cosmological inertial reference frame

Since our co-moving reference frame is freely falling in the Universe's overall gravitational field, it should be an inertial frame in which the galaxies would be freely falling with equations of motion that are free of accelerations. From Equation (5.9) these geodesic paths would satisfy

$$\frac{d^2x^\lambda}{d\tau^2} = -\Gamma^\lambda_{\mu\nu} \frac{dx^\mu}{d\tau} \frac{dx^\nu}{d\tau} = 0 \, .$$

In a co-moving system the spatial coordinates are unchanging, $dx^i = 0$ for $i = 1, 2, 3$; so only the $(\lambda; 0, 0)$ components of the affine connection are of importance and we must thus have

$$\Gamma^\lambda_{00} \left(\frac{dx^0}{d\tau} \right)^2 = 0$$

for all indices λ in order that $d^2x^\lambda/d\tau^2 = 0$ as desired. It thus suffices to have $\Gamma^\lambda_{00} = 0$ for all indices λ; it is left as an exercise for the student to show that this will be the case for a diagonal metric whenever g_{00} is a constant, which we thus take to be -1 in conformance with the SR invariant space-time interval.

8.1.4 Robertson–Walker metric

The resulting metric casts the expanding Universe of galaxies into the form of an inertial reference frame and is the basis for homogeneous, isotropic models of the expanding Universe. This is commonly called the **Robertson–Walker (RW)** metric of cosmology:[1]

Robertson–Walker metric

$$ds^2 = -c^2dt^2 + a(t)^2 \left[\frac{dr^2}{1 - K_0 r^2} + r^2 d\theta^2 + r^2 \sin^2\theta \, d\phi^2 \right] . \qquad (8.3)$$

The RW metric tensor is thus a diagonal one with components

$$\begin{aligned} g_{00} &= g_{tt} = -1 && \left(x^0 \equiv ct\right) , \\ g_{11} &= g_{rr} = \frac{a(t)^2}{1 - K_0 r^2} && \left(x^1 \equiv r\right) , \\ g_{22} &= g_{\theta\theta} = a(t)^2 r^2 && \left(x^2 \equiv \theta\right) , \\ g_{33} &= g_{\phi\phi} = a(t)^2 r^2 \sin^2\theta && \left(x^3 \equiv \phi\right) , \end{aligned} \qquad (8.4)$$

where the current curvature $K_0 = \pm 1/\left|R_0^2\right|$ and R_0 is the current radius of curvature.[2] The universal expansion function $a(t)$ describes the spatial evolution of the Universe and by convention is normalized to $a(t_0) = 1$, where t_0 is the current cosmic time. Cosmological models thus consist almost entirely of a functional form for $a(t)$ determined by solving the Einstein Field Equations with Equation (8.4) as the metric; plus the curvature K_0.

The coordinate or cosmic time t is well defined in terms of time-like hypersurfaces normal to the galaxy world lines bundle as illustrated in Figure 8.1. Each galaxy (or fundamental observer) can measure the same value of t by, e.g., noting the temperature of the cosmological background radiation (CMB) which, as we shall see several chapters hence, is a monotonic function of coordinate time. There are thus none of the usual SR complications of differing time systems for observers moving with respect to each other, and the cosmological time as measured locally conforms to the time kept on your watch. However, this choice of time and coordinate system can confuse the issues of distance and velocity; as we shall see in the following chapters, these concepts must be defined and interpreted with care in a co-moving setting.

Note, one more time, that the coordinates (r, θ, ϕ) of co-moving galaxies ('fundamental observers') do not change as a consequence of universal expansion and thus co-moving galaxies are, to first order, at rest with respect to all other

[1] Often referred to as the Friedmann–Robertson–Walker or the Friedmann–Lemaître–Robertson–Walker (FLRW) metric. Its currently accepted form is largely due to the work of cosmologist H.R. Robertson in the 1920s.

[2] R_0 is conveniently defined to be imaginary for negative curvatures; see Equation (9.49).

galaxies. Of course, galaxies *do* move around – M31 is approaching us at ~ 100 km/sec, for instance – but such 'peculiar' velocities arising from local gravitational effects are insignificant on cosmological scales in an expanding Universe, where the kinematics of universal expansion dominate over local effects.

8.1.5 Alternative metrics

As discussed in Section 6.2, different choices of coordinate systems can contain the same mathematical and physical properties, so we have great latitude in choosing the system with which to describe the expansion of a homogeneous and isotropic Universe. Two such alternatives to the system underlying the Robertson-Walker metric of Equation (8.3) are commonly used. The simpler of them employs a non-dimensional form of the radial coordinate, so that $r \to r/R$ and thus

$$ds^2 = -c^2dt^2 + R(t)^2 \left[\frac{dr^2}{1 - kr^2} + r^2 d\theta^2 + r^2 \sin^2\theta \, d\phi^2 \right], \tag{8.5}$$

where $k = KR^2$, the *curvature index*, is $(+1, 0, -1)$ for (positive, zero, negative) curvature, respectively. Another common choice is based on the coordinates underlying Equation (6.8) and leads to the metric

$$ds^2 = -c^2dt^2 + dr^2 + S_k(r)^2 \left[d\theta^2 + \sin^2(\theta) \, d\phi^2 \right], \tag{8.6}$$

where $S_k(r) = R \sin(r/R)$ for positive curvature, $R \sinh(r/|R|)$ for negative curvature, and r for flat geometries. Any one of these coordinate systems and accompanying metrics (as well as many others) will serve to describe the expansion of homogeneous and isotropic universes, although the intermediate computations will differ in form from one coordinate system to the next. As in several other facets of GR it is best to pick your system and stick with it. We will consistently use the system underlying the Robertson–Walker metric of Equation (8.3).

8.2 Field Equations of Cosmology

The Field Equations of Cosmology, applied to the RW metric of homogeneous and isotropic universes, are essentially differential equations for $a(t)$ and associated quantities as derived from the Einstein Field Equations of gravitation, Equation (7.8):

$$\mathcal{R}_{\mu\nu} = -\frac{8\pi G}{c^4} \left(\mathcal{T}_{\mu\nu} - \frac{1}{2} g_{\mu\nu} \mathcal{T} \right) - g_{\mu\nu} \Lambda ,$$

where the energy-momentum tensor $\mathcal{T}$ describes the density distribution of sources of gravitation, and the Ricci Tensor $\mathcal{R}_{\mu\nu}$ incorporates the effects of curvature in terms of the chosen metric. To find solutions to these equations in terms of $a(t)$ we first need to develop appropriate cosmological forms for $\mathcal{T}_{\mu\nu}$ and $\mathcal{R}_{\mu\nu}$.

8.2.1 Energy-momentum tensor

The basic model for energy/matter distribution in the Universe is that of radiation and of non-interacting galaxies, commonly called the 'dusty Universe'. The corresponding tensor representation for matter/energy densities is that for a perfect fluid, one with no shear or heat conduction. The SR form of the energy-momentum tensor for a perfect fluid is

$$\mathcal{T}^{\alpha\beta} = (\varepsilon + P)\, U^{\alpha} U^{\beta} + P\eta^{\alpha\beta} ,$$

where $U^{\mu} \equiv dx^{\mu}/d\tau$ is the fluid particle 4-velocity; $\varepsilon = \rho c^2$ is the energy/matter density; P is the pressure; and $\boldsymbol{\eta} = \mathrm{diag}(-1, 1, 1, 1)$ is the Minkowski metric of flat space-time. The appropriate generalization to curved space-time replaces $\boldsymbol{\eta}$ with the metric tensor $\mathbf{g}$ and, for conformance with the usual covariant form of the Einstein Tensor, we lower indices twice to obtain

$$\mathcal{T}_{\lambda\kappa} = g_{\alpha\kappa} g_{\beta\lambda} \mathcal{T}^{\alpha\beta} = (\varepsilon + P)\, U_{\kappa} U_{\lambda} + P g_{\lambda\kappa} .$$

In a co-moving coordinate system all spatial velocities are zero and $U_{\mu} = \mathrm{diag}(-1, 0, 0, 0)$, so

$$\begin{aligned} \mathcal{T}_{00} &= (\varepsilon + P)\,(-1)^2 - P &&= \varepsilon, \\ \mathcal{T}_{ii} &= (\varepsilon + P)\,(0)^2 + P g_{ii} &&= P g_{ii}, \end{aligned}$$

and thus the covariant energy-momentum density tensor for homogenous and isotropic cosmologies is just the simple diagonal tensor

$$\mathcal{T}_{..} = \mathrm{diag}\,(\varepsilon, P g_{ii}) \tag{8.7}$$

$(i = 1, 2, 3)$. Note that the diagonal nature of the energy-momentum tensor implies that the Ricci Tensor is also diagonal.

Cosmological Constant (II)

Returning now to the unreduced form of the Field Equations, Equation (7.7): we note that since $g_{00} = -1$ is dimensionless and $\mathcal{T}_{00} = \varepsilon$, Λ must carry the same dimensions as $\left(8\pi G/c^4\right)\varepsilon$, or of $(\text{length})^{-2}$; and the quantity

$$\varepsilon_{\Lambda} \equiv \frac{c^4}{8\pi G}\Lambda \tag{8.8}$$

carries dimensions of energy density and may be viewed as the energy density of the cosmological constant. Then the EFE may be written as

$$\mathcal{R}_{\mu\nu} - \frac{1}{2} g_{\mu\nu} \mathcal{R} = -\frac{8\pi G}{c^4} \left(\mathcal{T}_{\mu\nu} - g_{\mu\nu} \varepsilon_\Lambda \right) ,$$

motivating the formal incorporation of the cosmological constant into $\mathcal{T}_{\mu\nu}$. The reduced form of the EFE with augmented energy-momentum tensor is thus

$$\mathcal{R}_{\mu\nu} = -\frac{8\pi G}{c^4} \left(\breve{\mathcal{T}}_{\mu\nu} - \frac{1}{2} g_{\mu\nu} \breve{\mathcal{T}} \right) , \tag{8.9}$$

where the components of $\breve{\mathcal{T}}_{..}$ are understood to be those of matter and radiation (Equation (8.7)), augmented by cosmological constant terms. Without loss of generality we may represent these additional terms as generalized energy densities and pressures, so that $\breve{\mathcal{T}}_{..}$ retains the form of Equation (8.7). This re-formulation of the energy-momentum tensor will greatly facilitate computation of cosmological solutions of the EFE.

8.2.2 Cosmological solutions

Cosmological solutions of the EFE are equations relating the two adjustable elements of the RW metric – expansion function $a(t)$ and curvature K_0 – to the mass/energy density of the Universe as expressed in the energy-momentum tensor $\mathcal{T}$. Solutions begin with cosmological forms for the tensors on both sides of the EFE.

Energy-momentum

The right-hand side of the field equations takes the following form for cosmology: with $\breve{\mathcal{T}}_{00} = \varepsilon$ and $\breve{\mathcal{T}}_{ii} = g_{ii}P$, the scalar $\breve{\mathcal{T}} = \sum_i g^{ii} \breve{\mathcal{T}}_{ii}$ is easily seen to be $3P - \varepsilon$, and thus

$$-\frac{8\pi G}{c^4} \left(\breve{\mathcal{T}}_{00} - \frac{1}{2} g_{00} \breve{\mathcal{T}} \right) = -\frac{1}{2} \frac{8\pi G}{c^4} (3P + \varepsilon) \quad \longrightarrow \mathcal{R}_{00} , \tag{8.10}$$

$$-\frac{8\pi G}{c^4} \left(\breve{\mathcal{T}}_{ii} - \frac{1}{2} g_{ii} \breve{\mathcal{T}} \right) = -\frac{1}{2} \frac{8\pi G}{c^4} (\varepsilon - P) \, g_{ii} \quad \longrightarrow \mathcal{R}_{ii} . \tag{8.11}$$

Contributors to ε and P are conventional sources of mass and energy, and the cosmological constant.

Curvature

To evaluate the left-hand side of the Field Equations we need to use the metric tensor components (Equation (8.4)) to express the Ricci Tensor components as functions of the expansion factor $a(t)$. The requisite chain of calculations is

$$g_{..} \longrightarrow \Gamma^{\cdot}_{..} \longrightarrow \mathcal{R}_{..} \, .$$

This can be a tedious business (cf. Equation (7.5)), even if made simpler by the diagonal nature of all the tensors involved. The easy part is calculation of the affine connection, which is greatly aided by the nature of the RW metric. First, for diagonal metrics in general,

$$\begin{aligned}
\Gamma^{\alpha}_{\mu\nu} &= 0 \quad \text{if } \alpha \neq \mu \neq \nu \neq \alpha \, , \\
\Gamma^{\alpha}_{\mu\mu} &= -\frac{1}{2}\frac{1}{g_{\alpha\alpha}}\frac{\partial g_{\mu\mu}}{\partial x^{\alpha}} \quad \text{if } \alpha \neq \mu \, , \\
\Gamma^{\alpha}_{\alpha\mu} &= \frac{1}{2}\frac{1}{g_{\alpha\alpha}}\frac{\partial g_{\alpha\alpha}}{\partial x^{\mu}} \quad \text{for all } \alpha, \mu
\end{aligned}$$

(no implied sums in any of these expressions; cf. Problem 1 of Chapter 5). For the RW metric in particular it is also the case that affine connections with two or more time indices (0) are identically zero. With these as guides one can readily compute that, e.g.,

$$\Gamma^{1}_{01} = \frac{1}{c}\frac{\dot{a}}{a} \, , \quad \Gamma^{1}_{22} = -r\left(1 - K_0 r^2\right) \, , \text{ etc.,}$$

for the RW metric; and similarly for the 11 remaining non-zero components.[3] Note that $\dot{a}$ in these expressions denotes da/dt, not da/dx^0.

The affine connection components can then be used in computation of the Ricci Tensor. This is by far the nastiest step in calculation of the field equations; cf. Equation (7.5) once again. For the RW metric it suffices to compute only the diagonal components, $\mathcal{R}_{\mu\mu}$. Expressed with summations made explicit, these are

$$\mathcal{R}_{\mu\mu} = \sum_{\lambda}\left[\frac{\partial \Gamma^{\lambda}_{\mu\lambda}}{\partial x^{\mu}} - \frac{\partial \Gamma^{\lambda}_{\mu\mu}}{\partial x^{\lambda}} + \sum_{\eta}\left(\Gamma^{\eta}_{\mu\lambda}\Gamma^{\lambda}_{\mu\eta} - \Gamma^{\eta}_{\mu\mu}\Gamma^{\lambda}_{\lambda\eta}\right)\right] .$$

Weinberg (2008) derives the results as[4]

$$\mathcal{R}_{00} = \frac{3}{c^2}\frac{\ddot{a}}{a} \, , \tag{8.12}$$

$$\mathcal{R}_{ii} = -\frac{2}{c^2}\left(\frac{K_0 c^2}{a^2} + \frac{\ddot{a}}{2a} + \frac{\dot{a}^2}{a^2}\right) g_{ii} \, , \ i = 1, 2, 3 \, , \tag{8.13}$$

where by $\dot{a}$ is meant da/dt, not da/dx^0; and similarly for $\ddot{a}$.

[3] See, e.g., Narlikar (1993, p. 106) or Hartle (2003, Appendix B) for a complete list of RW affine connections.

[4] Weinberg's Equations 1.5.8 and 1.5.13, respectively.

8.2.3 Equations of expansion

Comparing Equations (8.10) and (8.12), and then Equations (8.11) and (8.13):

$$\frac{\ddot{a}}{a} = -\frac{4\pi G}{3c^2}(3P + \varepsilon) \ , \tag{8.14}$$

$$\frac{K_0 c^2}{a^2} + \frac{\ddot{a}}{2a} + \frac{\dot{a}^2}{a^2} = \frac{2\pi G}{c^2}(\varepsilon - P) \ .$$

Using the first of these to eliminate $\ddot{a}$ from the second,

$$\frac{K_0 c^2}{a^2} + \frac{\dot{a}^2}{a^2} = \frac{4\pi G}{c^2}(2\varepsilon) \ . \tag{8.15}$$

Equations (8.14) and (8.15), suitably arranged, are the basic forms of the **Friedmann Equations**[5] that constitute the basic solution to the Einstein Field Equations of cosmology:

$$\frac{\dot{a}^2}{a^2} = \frac{8\pi G}{3c^2}\varepsilon - \frac{K_0 c^2}{a^2} \ , \tag{8.16}$$

$$\frac{\ddot{a}}{a} = -\frac{4\pi G}{3c^2}(\varepsilon + 3P) \ . \tag{8.17}$$

Note that the first of these closely resembles the expansion equation of Newtonian cosmology, Equation (1.3), with curvature playing the role of total energy. But its interpretation, as developed in the following chapters, goes well beyond Newtonian physics.

8.3 Energy densities

Before we set out to solve the Friedmann Equations for the desired $a(t)$, we note that they are two equations in three unknown functions of time: $a(t)$, $\varepsilon(t)$, and $P(t)$. To solve them requires some additional information relating these functions, so we take a brief detour in this section to develop relations between ε, P, and a that allow us to write the Friedmann Equations entirely in terms of a. We start by identifying the components of energy density.

[5] Also called the Friedmann–Lemaître Equations, or the Friedmann–Lemaître–Robertson–Walker (FLRW) Equations.

8.3.1 Energy components

To the best of our current knowledge, the forms of mass/energy densities contributing to gravitation in a cosmological context are these.

Radiation The radiation energy density of the Universe, denoted $\varepsilon_{\rm r}$, is dominated by isotropic thermal radiation and primordial neutrinos, both remnants of the hot, early stages of universal evolution. The photon content of this radiation is well described by a black-body spectrum and is commonly referred to as the Cosmological Microwave Background Radiation, aka 'CMB'. The neutrino component is hypothesized but apparently well-understood, even if unobserved (the mean energy of primordial neutrinos is much too low for detection). The energy density of radiation from stars, etc., is lower than that of the CMB by orders of magnitude and can safely be ignored for large-scale cosmological purposes.

Matter The energy density of matter, $\varepsilon_{\rm m} = \rho_{\rm m}c^2$, is dominated by non-relativistic baryonic and non-baryonic ('dark matter') components. Relativistic matter is either in the form of low-mass neutrinos and is included in the radiation budget, or existed so early in the Universe that its influence on expansion was overwhelmed by that of pure radiation and so can safely be ignored.

Cosmological Constant From Equation (8.8) an energy density corresponding to the cosmological constant may be formally defined as

$$\varepsilon_\Lambda \equiv \frac{c^4}{8\pi G}\Lambda \,. \tag{8.18}$$

The cosmological constant energy density is commonly referred to as '**dark energy**' although, as we shall see in Chapter 13, that term properly includes more possibilities than just the Cosmological Constant. But for the remainder of the discussion of universal expansion, the terms 'cosmological constant energy' and 'dark energy' are effectively equivalent.

With these definitions the Cosmological Field Equation may be written as

$$\frac{\dot a^2}{a^2} = \frac{8\pi G}{3c^2}\left(\varepsilon_{\rm r} + \varepsilon_{\rm m} + \varepsilon_\Lambda\right) - \frac{K_0 c^2}{a^2} \,, \tag{8.19}$$

$$\frac{\ddot a}{a} = -\frac{4\pi G}{3c^2}\left(\varepsilon_{\rm r} + \varepsilon_{\rm m} + \varepsilon_\Lambda + 3P\right) \,, \tag{8.20}$$

where the εs are implicitly functions of t (or a), and the P in the second equation contains contributions from radiation and dark energy (the galaxies essentially constitute a collisionless, and thus pressure-free, gas).

8.3.2 Energy Equation

An equation relating mass/energy density evolution to the expansion function can be derived directly from the divergence-free property of the energy-momentum tensor, or from the two Friedmann Equations themselves; but it is instructive to derive it instead from fundamental thermodynamic principles. The First Law of Thermodynamics reads

$$dE = Q - PdV + \mu dN\ ,$$

where E is the total energy, P and V pressure and volume, Q the heat flow, N the number of particles, and μ the chemical potential. Except for the very earliest times in the Universe's history, Q is effectively zero and N is a constant except for photons for which $\mu = 0$; so (dividing by dt),

$$\dot{E} + P\dot{V} = 0\ . \tag{8.21}$$

Now consider the case of a small sphere of co-moving (coordinate) radius r_{s}, so its proper radius[6] is $r_{\mathrm{p}} = r_{\mathrm{s}} a\,(t)$ and its proper volume is

$$V = \frac{4}{3}\pi r_{\mathrm{p}}^3 = \frac{4}{3}\pi r_{\mathrm{s}}^3 a^3\ .$$

Then

$$\dot{V} = 4\pi r_{\mathrm{s}}^3 a^2 \dot{a} = 3\frac{\dot{a}}{a} V\ . \tag{8.22}$$

The total energy in the sphere is $E = V\varepsilon$, where ε is the energy density, so

$$\dot{E} = \dot{V}\varepsilon + V\dot{\varepsilon} = V\left(\dot{\varepsilon} + 3\frac{\dot{a}}{a}\varepsilon\right)\ . \tag{8.23}$$

Inserting the last two equations into Equation (8.21),

$$V\left(\dot{\varepsilon} + 3\frac{\dot{a}}{a}\varepsilon + 3\frac{\dot{a}}{a}P\right) = 0\ .$$

Dividing by V and rearranging terms yields the **Energy Equation** of cosmology:

Energy Equation

$$\dot{\varepsilon} + 3\frac{\dot{a}}{a}\left(\varepsilon + P\right) = 0\ . \tag{8.24}$$

[6] These formulae for proper radius and volume stem from the more general formulae derived in Section 9.3, in the limit of small coordinate increments.

This equation is on an equal footing with the two Friedmann Equations; indeed, it can be derived from them (as Problem 2 at the end of this chapter invites you to prove) and so does not constitute a third independent differential equation for (a, ε, P), but will nonetheless prove to be very useful.

8.3.3 Equation of State

We can relate pressure to energy density, and thus eliminate it as an independent variable, by employing an Equation of State (EOS). Generically, an EOS relates pressure to energy density in the form

$$P = w\varepsilon \; , \tag{8.25}$$

where w is (hopefully) a constant of known value. The values of w for the known energy components of the Universe are as follows.

- For **black-body radiation** $P_\mathrm{r} = \varepsilon_\mathrm{r}/3$, a standard thermodynamic relation that also applies to highly relativistic matter.
- For **non-relativistic matter** we can approximate the pressure by the Ideal Gas Law, $P_\mathrm{m} = (\rho/\bar{m})\, k_\mathrm{B} T$, where $\bar{m}$ is the mean particle mass and k_B is Boltzmann's constant. Since $\rho = \varepsilon/c^2$ this can be written

$$P_\mathrm{m} = \frac{k_\mathrm{B} T}{\bar{m} c^2} \varepsilon_\mathrm{m} = \frac{E_\mathrm{thermal}}{E_\mathrm{rest}} \varepsilon_\mathrm{m} \; ,$$

  so $w \approx 0$ for non-relativistic matter for which the mean thermal energy $k_\mathrm{B} T$ is much less than the rest energy $\bar{m} c^2$.
- For **dark energy** arising from the Cosmological Constant the situation is more interesting. First, from Equation (8.18) we note that ε_Λ is a constant if Λ is (and which it is assumed to be for current purposes), so $\dot{\varepsilon}_\Lambda = 0$. Then from the Energy Equation $P_\Lambda = -\varepsilon_\Lambda$ and $w_\Lambda = -1$: the pressure arising from the cosmological constant is negative, so that Λ corresponds to a force of expansion.

The origin of this negative pressure in a constant energy density can be illustrated by the following *Gedankenexperiment*. Consider a 'space' of ordinary vacuum of zero energy density, in which we embed a cylinder of volume V filled with 'Λ-space' of constant energy density ε, so its energy is $E = \varepsilon V$; and with pressure P_Λ. Pull the piston blocking the cylinder entrance out slowly, so the cylinder volume increases: $V \rightarrow V + \Delta V$. By the first law of thermodynamics the energy in the cylinder changes by $\Delta E = -P_\Lambda \Delta V$. But this is also $\Delta E = \varepsilon_\Lambda \Delta V$ if ε_Λ

remains constant, so we must have $P_\Lambda = -\varepsilon_\Lambda$ as implied by the energy equation, and thus $w_\Lambda = -1$. A universal component with constant energy density thus implies a negative pressure and – eventually – accelerating expansion (as we shall see). A negative pressure is thus a pressure *from* a system *on* its surroundings, corresponding to forces of expansion.

To summarize: the EOS parameters for the standard energy components are

$$w = \begin{cases} 1/3\,, & \text{radiation,} \\ 0\,, & \text{non-relativistic matter,} \\ -1\,, & \Lambda\text{-dark energy.} \end{cases} \tag{8.26}$$

The term $\varepsilon + 3P$ in the second Friedmann Equation (8.17) thus can be written $\varepsilon\,(1 + 3w)$ to eliminate P as a variable, so that the Friedmann Equations become a system of two equations that can be solved simultaneously for both a and ε as functions of time (for given gravitating contents of the Universe). But we can go even further to find useful expressions relating the evolution of ε (or ρ) to that of a in a co-moving cosmological model.

8.3.4 Energy density evolution

For gravitating components characterized by an equation of state the Energy Equation can be used to directly relate mass and energy densities to the expansion function $a\,(t)$ and thus provide the necessary additional constraint to allow the Friedmann Equations to be fully solved. Using the EOS parameter to eliminate P,

$$\dot{\varepsilon} + 3\frac{\dot{a}}{a}\varepsilon\,(1 + w) = 0\,.$$

Dividing through by ε/dt changes the independent variable from t to a and yields an alternative form of the Energy Equation:

$$\begin{aligned} \frac{d\varepsilon}{\varepsilon} &= -3\,[1 + w\,(a)]\,\frac{da}{a}\,, \\ \Rightarrow \varepsilon\,(a) &= \varepsilon_0 \exp\left[-3\int_1^a \frac{1 + w\,(\alpha)}{\alpha}d\alpha\right], \end{aligned} \tag{8.27}$$

where ε_0 is the current value. If w is a constant – as is the case with 'normal' sources of gravitation – this equation reduces to

$$\varepsilon\,(a) = \varepsilon_0\,a^{-3(1+w)} \qquad \text{(constant } w\text{).} \tag{8.28}$$

For the presumed contents of the Universe, w is the constant listed in (8.26) and the corresponding energy densities thus vary with the universal expansion

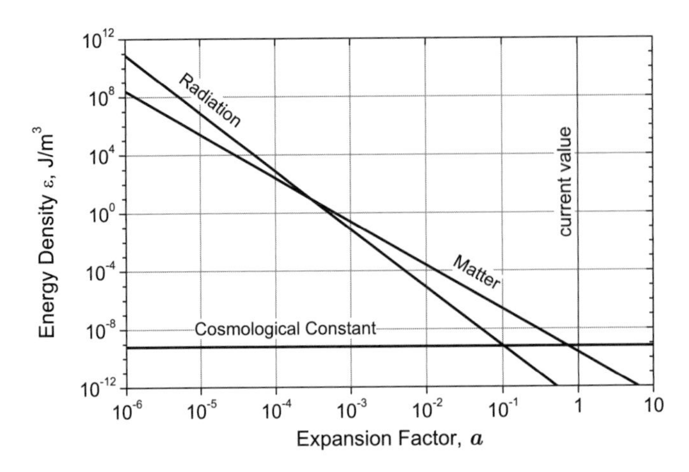

Figure 8.2 Evolution of energy densities for the CCM current densities (see Chapter 15).

function as

$$\text{radiation:} \quad \varepsilon_{\mathrm{r}}(a) = \varepsilon_{\mathrm{r},0}\, a^{-4}, \tag{8.29}$$

$$\text{matter:} \quad \varepsilon_{\mathrm{m}}(a) = \varepsilon_{\mathrm{m},0}\, a^{-3}, \tag{8.30}$$

$$\text{Cosmological Constant:} \quad \varepsilon_{\Lambda}(a) = \varepsilon_{\Lambda,0}, \tag{8.31}$$

where $\varepsilon_{\mathrm{r},0}$, $\varepsilon_{\mathrm{m},0}$, and $\varepsilon_{\Lambda,0}$ are the energy densities in the current epoch.

That ε_{m} and ε_{r} differ by a factor of a is a consequence of the redshift in an expanding Universe: since $\lambda \propto a$ and $E_{\mathrm{photon}} \propto \lambda^{-1}$, the energy density of radiation suffers an additional diminution due to redshift. Note that if the Universe started off being dominated by radiation (as we believe to be the case), the differing forms of energy evolution for the three components imply that, after some later time, the dominant form of energy will be that of the Cosmological Constant, with (possibly) matter dominating at intermediate times. Exactly when this happens depends upon the time evolution of the scale factor a, which is model-dependent. Figure 8.2 demonstrates such evolution for our current best estimate of energy densities in the Universe, that of the CCM (Chapter 15). Radiation yields to matter as the dominant gravitational source near $a \approx 3 \times 10^{-4}$, and the cosmological constant assumes the dominant role just before the current time ($a = 1$).

8.4 Friedmann Equations

With the help of these energy density and EOS relations we can re-write the Friedmann Equations in terms of the expansion function alone.

Friedmann Equations

$$\textbf{expansion:} \quad \frac{\dot{a}^2}{a^2} = \frac{8\pi G}{3c^2}\left(\frac{\varepsilon_{r,0}}{a^4} + \frac{\varepsilon_{m,0}}{a^3} + \varepsilon_{\Lambda,0}\right) - \frac{K_0 c^2}{a^2}\,, \tag{8.32}$$

$$\textbf{acceleration:} \quad \frac{\ddot{a}}{a} = -\frac{4\pi G}{3c^2}\left(2\frac{\varepsilon_{r,0}}{a^4} + \frac{\varepsilon_{m,0}}{a^3} - 2\varepsilon_{\Lambda,0}\right)\,. \tag{8.33}$$

Equations (8.32) and (8.33) are, in effect, the Einstein Field Equations in a cosmological context. The two equations are equivalent despite their different forms and names; either may be used to solve for the expansion function $a(t)$, but the first is more often used for that purpose because of its lower order. The relative simplicity of these equations – in comparison to the Einstein Field Equations, from which they derive (cf. Equation (7.5)) – is largely a consequence of the symmetry implied by the presumed homogeneity and isotropy of space, and of a clever choice of coordinate system.

8.4.1 Cosmological Constant (III)

The Cosmological Constant plays a unique role in the range of possible solutions to the Friedmann Equations. For positive $(\Lambda,\ \varepsilon_\Lambda)$ and sufficiently large a, the acceleration becomes positive (Equation (8.33)) and $\dot{a} \propto a$ (Equation (8.32)), which is the differential equation for exponential growth.

> *A positive cosmological equation leads eventually to exponential expansion of the Universe*;

and thus, of course, to an accelerating expansion. The cosmological constant was introduced by Einstein for a somewhat different reason: to ensure the possibility of a static, non-expanding Universe. The appropriate portion of the Acceleration Equation,

$$\ddot{a} \rightarrow a\frac{8\pi G}{3c^2}\varepsilon_\Lambda = a\frac{c^2\Lambda}{3}\,,$$

implies a force of repulsion proportional to distance;[7] in this sense the constant acts to oppose gravity, which is why Einstein chose it to balance forces in his static Universe. But since the distance dependency of this repulsive force ($\propto d$)

[7] The distance between galaxies is proportional to a; see Equation (I.1) or, more generally, Section 9.2.

differs from that of the attractive force of gravitation $\left(\propto d^{-2}\right)$, a static solution to the modified Field Equations is inherently unstable: if a galaxy moves a bit from its equilibrium position the forces on it will become unbalanced so as to move the galaxy even further, and the whole static assembly will either fly apart or come crashing together. It's a bit like balancing a pencil on its tip: you can do it (in principle), but tilt the pencil a bit and it will fall over. Einstein's static model – for which he invented the cosmological constant – is not realizable.

Problem 5 invites you to show that a positive cosmological constant is required for a static model, so Einstein introduced the constant into his Field Equations for just this purpose, employing the reasoning outlined here but without appreciating its inherent instability (later pointed out to him by Eddington). Einstein only revoked his proposal when observations showed that the Universe was expanding. The cosmological constant was re-introduced in the late 1990s in order to explain the faintness of distant standard candles which seemingly required acceleration of universal expansion (see Chapter 13). It is a great pity that Einstein did not live to see his original proposal resurrected.

8.5 Simple expansion models

Models for expansion of the Universe are solutions $a(t)$ to the Friedmann Equations; from Equation (8.32) these will constitute a four-dimensional family of expansion functions parameterized by the energy densities and curvature. Such solutions will be examined in detail in the following chapters, based on the supporting material developed in Chapters 9 and 10. As a sneak preview of cosmological models, here are two particularly simple examples of historical interest.

The **de Sitter model** was the first successful model describing an expanding universe, proposed by Willem de Sitter in 1917. It is characterized (in modern terms) as a geometrically flat model containing only the Cosmological Constant: $\varepsilon_{r,0} = \varepsilon_{m,0} = K_0 = 0$. For this parameterization the Expansion Equation (8.32) is

$$\dot{a} = \left(\frac{8\pi G}{3c^2}\varepsilon_{\Lambda,0}\right)^{1/2} a\,,$$

to which the solution is

$$a(t) = \left[\exp\left(\frac{8\pi G}{3c^2}\varepsilon_{\Lambda,0}\right)^{1/2}\right](t - t_0) \qquad \text{(de Sitter).} \tag{8.34}$$

The de Sitter model is an exponentially expanding one.[8]

[8] Exponentially expanding when expressed in terms of co-moving coordinates and the RW metric; de Sitter's original model employed a different metric and was, formally, static.

The **Einstein–de Sitter model** was proposed jointly by Einstein and de Sitter in 1932, following Hubble's discovery of universal expansion. It is characterized as a geometrically flat model containing only matter: $\varepsilon_{r,0} = \varepsilon_{\Lambda,0} = K_0 = 0$. The corresponding expansion equation is

$$\dot{a} = \left(\frac{8\pi G}{3c^2}\varepsilon_{m,0}\right)^{1/2} a^{-1/2}\,, \quad \text{(Einstein–de Sitter)}$$

to which the solution with $a\,(0) = 0$ is

$$a\,(t) = \left(\frac{6\pi G}{c^2}\varepsilon_{m,0}\right)^{1/3} t^{2/3}\,. \tag{8.35}$$

This **flat, matter-only** model will be frequently employed as an example model in following chapters.

8.6 Reality check: expanding space

An even simpler (if unrealistic) model is that for a totally empty Universe, for which the expansion equation is

$$\dot{a}^2 = -K_0 c^2\,.$$

The trivial solution is $K_0 = \dot{a} = 0$: a flat, static Universe. But the expansion equation is also satisfied by an expanding Universe with negative curvature:

$$a\,(t) = \sqrt{-K_0}\,ct\,,$$

with $K_0 < 0$. This raises two questions: if the Universe is truly empty, what is it whose expansion is described by $a\,(t)$, and what is it that is curved? The canonical answer to such questions is: space itself. But in this case the apparent expansion clearly arises entirely from the choice of a co-moving coordinate system, so that the coordinate system is expanding and the accompanying metric is curved. As it stands, this has nothing to do with anything material or substantive and, while the issue is considerably more nuanced when matter/energy is present, it is apparent from this example that expansion and curvature can be, in part, artifacts of the choice of coordinates and metric.

The reality of 'curved space-time' is part of the philosophical foundation of GR,[9] but in cosmological applications many phenomena can be understood without invoking curvature and expansion of space itself: the expanding Universe of galaxies can usefully be thought of in terms of galaxies falling upward in the

[9] See Section 17 of Graves (1971) for arguments to this effect.

Universe's overall gravitational field, rather than as expanding space. Whether you choose to view cosmological expansion and curvature as properties of space itself, or as descriptions of the Universe of galaxies expressed in terms of the chosen metric, is largely immaterial to analysis of cosmological observations. But the 'expanding space' picture, as intuitively compelling as it is, can lead to conceptual difficulties (as will be revealed in the following chapters), and it is useful to keep in mind that some of the properties commonly ascribed to space itself are consequences of the chosen metric and, in that sense, are not fundamental properties of space.

Problems

1. Use Equation (5.10) for the affine connection to show that $\Gamma^{\lambda}_{00} = 0$ if g_{00} is a constant and g is diagonal, as claimed in Section 8.1.2.
2. Use the Friedmann Equations (8.16) and (8.17) to derive the Energy Equation (8.24) and thus show that these three equations are not independent.
3. Assuming that the current energy density of the Universe ($\sim 10^{-9}$ J/m^3) is due entirely to the Cosmological Constant, compute its value and the corresponding scale length $R_\Lambda = \Lambda^{-1/2}$. Compare this last value to the current 'size' of the visible Universe of $\sim$ 10 Gpc.
4. Show that the Acceleration Equation (8.33) can be derived directly from the Expansion Equation (8.32), and thus that the two are equivalent (given proper initial conditions).
5. Einstein's original solution to the cosmological field equations presumed a static Universe ($\dot{a} = \ddot{a} = 0$) containing only matter, in conformance with observational evidence at that time ($\sim$ 1915).

   (a) Use the Friedmann Equations to show that the static presumption requires a positive cosmological constant and calculate its value if the mass density is $\sim 10^{-27}$ kg/m^3.
   (b) Show that such a Universe must be closed and find an estimate for its radius of curvature, $R_0 = K_0^{-1/2}$. Compare this value to the observational limits for the size of the Universe of that time, of several Mpc.

6. If an energy component has an EOS parameter that varies with expansion as $w(a) = w_0 a$, where w_0 is the current value:

   (a) find an expression for $\varepsilon(a)$ in terms of ε_0 and w_0;
   (b) graphically display this function for various values of w_0 (including negative ones) alongside the standard $\varepsilon_0 a^{-3}$ for matter (corresponding to $w_0 = 0$). Comment on the likely course of evolution of a model containing such an energy component at large values of a.

9

Cosmography

The solutions of the Friedmann Equations constitute a four-parameter family of expansion functions $a(t)$, whose members are characterized by values of the three energy densities $\varepsilon_{x,0}$ and the curvature constant K_0. It is a major goal of modern cosmology to determine the Universe's curvature and the form of the expansion function from observations. To do so requires development of relations between observable quantities such as redshift, brightness, apparent size, etc.; and the expansion function parameters. This is the realm of *Cosmography,* a term commonly meant to encompass the study of the Universe at large without reference to a specific model; i.e., in which the expansion function $a(t)$ is unspecified.

This rather long chapter develops the tools needed to describe the Universe's expansion in terms of observations and cosmological parameters. Section 9.1 introduces alternative parameters for use in the Friedmann Equations, that are more readily related to observations. Section 9.2 derives the connections between model parameters and the basic properties of expansion: time, distance, redshift. Section 9.3 uses these to derive the observable consequences of expansion for comparison with observations. These will be used in later chapters to characterize cosmological models.

9.1 Model parameters

9.1.1 Expansion (Hubble) Parameter

The relative expansion rate $\dot{a}/a$ occurs so frequently in cosmology that it has been enshrined as the **Hubble Parameter**,

$$\text{Hubble Parameter:} \quad H \equiv \frac{\dot{a}}{a} \,. \tag{9.1}$$

The dimensions of H are those of inverse time. Associated with the Hubble Parameter are characteristic expansion times and distances,

$$\text{Hubble Time:} \quad t_{\rm H} \equiv 1/H, \tag{9.2}$$

$$\text{Hubble Distance:} \quad d_{\rm H} \equiv ct_{\rm H} = c/H. \tag{9.3}$$

The Hubble Time can usefully be thought of as (roughly) the time required for the Universe to double in size.

The current value of H is the **Hubble Constant**,

$$\text{Hubble Constant:} \quad H_0 \equiv \left(\frac{\dot{a}}{a}\right)_{t=t_0} = \dot{a}\,(t_0)\ . \tag{9.4}$$

It is a fundamental observable characteristic of the expanding Universe. From Equation (3.4) the Hubble Constant satisfies

$$H_0 \approx c\frac{z}{d} \quad (\text{for } z \ll 1)\ ,$$

where z and d are galaxy redshift and distance, respectively, and are both observable; so that H_0 may be estimated from observed distances and redshifts to nearby galaxies. This aspect of the Hubble Parameter is crucial to our understanding of the Universe and will be further examined later in this chapter.

The currently adopted (by most people, approximately) observational estimates for H_0 in different units are:

$$H_0 \approx \begin{cases} 0.074 & \text{Gyr}^{-1}, \\ 2.3 \times 10^{-18} & \text{sec}^{-1}, \\ 72 & \text{km/sec/Mpc}, \end{cases}$$

all plus/minus $\sim$ 5% (current estimate, subject to change!). The last of these, in mixed units, derives from the early days of observational cosmology when galaxy redshifts were interpreted directly as Doppler shifts: $V = H_0 d$. These mixed units are the most often employed by observational cosmologists, but are a nuisance in computing densities and other physical quantities; their conversion to pure time units is facilitated by

$$H_0 \begin{bmatrix} \text{in Gyr}^{-1} \\ \text{in sec}^{-1} \end{bmatrix} = \left.\begin{matrix} 1.02 \times 10^{-3} \\ 3.24 \times 10^{-20} \end{matrix}\right\} \times H_0 \left[\text{in km/sec/Mpc}\right]\ . \tag{9.5}$$

Note that the canonical $H_0 = 72$ km/sec/Mpc $= 0.074$ Gyr^{-1} corresponds to Hubble Time of $t_{\rm H} \approx 13.5$ Gyr, which sets the scale (if not the exact value) for the age of the expansion.

The history of modern cosmology is one in which our estimates for H_0 have changed dramatically and frequently, as the quality of cosmological observations evolved. Hubble's original estimate was $\sim$ 500 km/sec/Mpc; over the course of the next 60 years the estimate drifted down to the vicinity of 100, and since the 1990s it has flopped around between $\sim$ 55 and 80 km/sec/Mpc (all with error estimates of a few percent). Since our interpretations of observations in cosmological terms typically depend sensitively on the chosen value of H_0, cosmologists frequently choose to present those interpretations in a scalable form in terms of $h \equiv H_0/(100 \text{ km/sec/Mpc})$, so that the current canonical value is $h \approx 0.72$.

9.1.2 Energy density parameters

We can write the Expansion Equation (8.16) in terms of the Hubble Parameter as

$$H^2 = \frac{8\pi G}{3c^2}\varepsilon - \frac{K_0 c^2}{a^2} , \tag{9.6}$$

where $\varepsilon \equiv \varepsilon_{\mathrm{r}} + \varepsilon_{\mathrm{m}} + \varepsilon_{\Lambda}$. If the Universe is to be flat ($K_0 = 0$) the combined energy density must then take on the special value

$$\varepsilon \to \varepsilon_{\mathrm{c}} \equiv \frac{3c^2 H^2}{8\pi G} . \tag{9.7}$$

This is the **Critical Energy Density**, that required of a flat Universe expanding at rate H. Its currently estimated value, for the canonical $H_0 = 72$ km/sec/Mpc $= 0.074 \text{ Gpc}^{-1}$, is

$$\begin{aligned} \varepsilon_{\mathrm{c},0} = \frac{3c^2 H_0^2}{8\pi G} &\approx 8.8 \times 10^{-10} \text{ J/m}^3 , \\ &\approx 5500 \text{ MeV/m}^3 . \end{aligned} \tag{9.8}$$

The corresponding mass density is

$$\rho_{\mathrm{c},0} = \varepsilon_{\mathrm{c},0}/c^2 \approx 9.8 \times 10^{-27} \text{ kg/m}^3 , \tag{9.9}$$

$$\approx 1.5 \times 10^{11} \text{ M}_{\odot}/\text{Mpc}^3 , \tag{9.10}$$

or the equivalent of about 6 baryons per cubic meter.

From the above equations $8\pi G/3c^2 = H^2/\varepsilon_{\mathrm{c}} = H_0^2/\varepsilon_{\mathrm{c},0}$, so we can write the Expansion Equation (8.32) as

$$\left(\frac{\dot{a}}{a}\right)^2 = \frac{H_0^2}{\varepsilon_{\mathrm{c},0}}\left(\frac{\varepsilon_{\mathrm{r},0}}{a^4} + \frac{\varepsilon_{\mathrm{m},0}}{a^3} + \varepsilon_{\Lambda,0}\right) - \frac{K_0 c^2}{a^2} .$$

The non-dimensional forms of energy densities are then defined as the **energy density parameters**; generically,

$$\Omega_x \equiv \frac{\varepsilon_x}{\varepsilon_{\mathrm{c}}} . \tag{9.11}$$

$$\Rightarrow \Omega_{x,0} = \frac{\varepsilon_{x,0}}{\varepsilon_{\mathrm{c},0}} = \frac{8\pi G}{3c^2 H_0^2}\varepsilon_{x,0} \qquad \text{for} \quad x = \mathrm{r, m}, \Lambda . \tag{9.12}$$

Note that $\Omega_x = 1$ if $\varepsilon_x = \varepsilon_c$. The Expansion Equation may then be written in terms of these parameters as

$$\left(\frac{\dot{a}}{a}\right)^2 = H_0^2\left(\frac{\Omega_{\mathrm{r},0}}{a^4} + \frac{\Omega_{\mathrm{m},0}}{a^3} + \Omega_{\Lambda,0}\right) - \frac{K_0 c^2}{a^2} \ . \tag{9.13}$$

The total or combined energy density parameter is denoted by Ω without a component subscript:

$$\Omega \equiv \Omega_{\mathrm{r}} + \Omega_{\mathrm{m}} + \Omega_{\Lambda} \ . \tag{9.14}$$

Its current value,

$$\Omega_0 = \Omega_{\mathrm{r},0} + \Omega_{\mathrm{m},0} + \Omega_{\Lambda,0} = \frac{1}{\varepsilon_{\mathrm{c},0}}\left(\varepsilon_{\mathrm{r},0} + \varepsilon_{\mathrm{m},0} + \varepsilon_{\Lambda,0}\right) \ , \tag{9.15}$$

will be of some use in the coming sections. Currently estimated values for the energy density parameters are $\Omega_{\mathrm{r},0} \approx 0$, $\Omega_{\mathrm{m},0} \approx 0.27$, $\Omega_{\Lambda,0} \approx 0.73$; so that $\Omega_0 \approx 1$. The combined energy densities as currently estimated suggest a nearly critical density to the Universe.

9.1.3 Geometric curvature

The final term in the Expansion Equation is directly related to curvature: $K = \pm|R|^{-2}$, where R is the radius of curvature in a uniformly curved space, and $|R| \propto a$ so that $K_0/a^2 = K$ is the geometric curvature,

$$\text{geometric curvature:} \quad K = \frac{K_0}{a^2} \ . \tag{9.16}$$

In cosmological application, spatial curvature will play a significant role only on scales comparable to, or larger than, $|R|$; on smaller scales space is effectively flat.

Cosmological curvature is closely related to energy densities: at the current time ($a = 1$) the Expansion Equation (9.13) is

$$\begin{aligned} H_0^2 &= H_0^2\left(\Omega_{\mathrm{r},0} + \Omega_{\mathrm{m},0} + \Omega_{\Lambda,0}\right) - K_0 c^2 \\ &= H_0^2 \Omega_0 - K_0 c^2 \ , \end{aligned}$$

from which

$$K_0 c^2 = H_0^2\left(\Omega_0 - 1\right) \ . \tag{9.17}$$

From this and Equation (9.16) the geometric curvature of the Universe evolves with expansion as

$$K = \frac{H_0^2}{c^2} \frac{\Omega_0 - 1}{a^2} , \tag{9.18}$$

so K carries the same sign as $\Omega_0 - 1$: $K\ (<, =, >)\ 0$ if $\Omega_0\ (<, =, >)\ 1$. We will thus often use $\Omega_0 - 1$, or just Ω_0, as a useful synonym for curvature. Note that, from the above equation, curvature decreases (in absolute value) monotonically, but does not change sign, as the Universe expands: open universes stay open, closed ones stay closed, and flat universes stay flat. The Universe is geometrically flat if, and only if, $\Omega_0 = 1$.

Since observations suggest $\Omega_0 \approx 1$, the geometry of the Universe is very nearly flat: $|\Omega_0 - 1| < 1$, and possibly much less than this, in all currently plausible models. From this and Equation (9.17),

$$\begin{aligned} |K_0| &\lesssim 0.06\ \mathrm{Gpc}^{-2} , \\ \Rightarrow |R_0| &\gtrsim 4\ \mathrm{Gpc} , \end{aligned} \tag{9.19}$$

if current estimates of mass/energy densities are correct (but see Chapters 12–14 for analyses of the reliability of these estimates). Note that relativistic effects are likely to be of importance only on scales that are a significant fraction of $|R_0|$.

9.1.4 Parameterized Friedmann Equations

We can use Equation (9.17) to eliminate curvature K from the Expansion Equation and thus write it entirely in terms of energy densities:

$$\left(\frac{\dot{a}}{a}\right)^2 = H_0^2 \left(\frac{\Omega_{\mathrm{r},0}}{a^4} + \frac{\Omega_{\mathrm{m},0}}{a^3} + \Omega_{\Lambda,0} - \frac{\Omega_0 - 1}{a^2} \right) . \tag{9.20}$$

The Friedmann Equations may now be written as ordinary differential equations containing only observable parameters:

expansion:
$$\left(\frac{da}{dt}\right)^2 = \dot{a}^2 = H_0^2 \left[\frac{\Omega_{\mathrm{r},0}}{a^2} + \frac{\Omega_{\mathrm{m},0}}{a} + a^2 \Omega_{\Lambda,0} - (\Omega_0 - 1) \right] , \tag{9.21}$$

acceleration:
$$\frac{d^2 a}{dt^2} = \ddot{a} = H_0^2 \left[-\frac{\Omega_{\mathrm{r},0}}{a^3} - \frac{\Omega_{\mathrm{m},0}}{2a^2} + a\, \Omega_{\Lambda,0} \right] , \tag{9.22}$$

with the understanding that $\Omega_0 = \Omega_{\mathrm{r},0} + \Omega_{\mathrm{m},0} + \Omega_{\Lambda,0}$. Note that a positive cosmological constant implies $\Omega_{\Lambda,0} > 0$, which, from the Acceleration Equation, opens the possibility of an accelerating ($\ddot{a} > 0$) expansion at sufficiently large values of a.

The solution set to the Friedmann Equations is a four-parameter family characterized by energy density parameters $\left(\Omega_{r,0}, \Omega_{m,0}, \Omega_{\Lambda,0}\right)$ and by the Hubble Constant, H_0. In the remainder of this chapter we develop some generic properties of these solutions, before studying specific examples in later chapters. As an illustrative example we will often invoke the Einstein–de Sitter model: flat, containing only matter; so that $\Omega_{m,0} = \Omega_0 = 1$ and $\Omega_{r,0} = \Omega_{\Lambda,0} = 0$. Then the Expansion Equation becomes $\dot{a} = H_0 a^{-1/2}$, to which the solution (with $a(0) = 0$) is

$$a(t) = \left(\frac{3}{2} H_0\, t\right)^{2/3}, \tag{9.23}$$

$$= (t/t_0)^{2/3}, \tag{9.24}$$

where $t_0 = 2/(3H_0)$ is the current time.

9.2 Expansion descriptors

A fundamental goal of cosmology is the determination of the form of the expansion function $a(t)$, and thereby the elucidation of the Universe's past and future. But neither a nor t are observable, so to proceed further we must develop observable proxies for them.

9.2.1 Redshift

A fundamental observable quantity of distant galaxies is the systematic scaling of their spectra: $\lambda_{\text{observed}} \to (z+1)\,\lambda_{\text{laboratory}}$, where z is the **redshift** of the spectrum, defined in terms of wavelength by

$$z \equiv \frac{\lambda_o - \lambda_e}{\lambda_e} = \frac{\lambda_o}{\lambda_e} - 1; \tag{9.25}$$

where e denotes as emitted from a distant galaxy and o denotes as observed by us. For objects in relative radial motion in an inertial reference frame this quantity is given simply by the usual Doppler shift formulae, but in a co-moving coordinate system galaxies are not moving with respect to each other, and in a cosmological context the redshift is partially gravitational in origin; so the relation of galaxy redshifts to observable quantities is more complicated than in classical mechanics.

To relate observed cosmological redshifts to universal expansion we thus need to express them in terms of the Robertson–Walker metric, which we repeat here:

$$ds^2 = -c^2 dt^2 + a(t)^2 \left[\frac{dr^2}{1 - K_0 r^2} + r^2 d\theta^2 + r^2 \sin^2(\theta)\, d\phi^2\right].$$

We choose a coordinate system with the origin at our Galaxy, radial coordinate r_g at the observed galaxy, and a line-of-sight to the galaxy along a purely radial direction, so $d\theta = d\phi = 0$. Since light travels along null geodesics we also have $ds = 0$, and thus

$$cdt = \pm a(t) \frac{dr}{\sqrt{1 - K_0 r^2}} .$$

Rearranging and integrating along the ray path,

$$\int_{t_e}^{t_r} \frac{dt}{a(t)} = \frac{1}{c} \int_0^{r_g} \frac{dr}{\sqrt{1 - K_0 r^2}} ,$$

where the photon was emitted at time t_e and received at time t_r. Since co-moving coordinates are unchanging for a given galaxy, the right-hand side of this equation (which is a function only of r_g) is the same no matter what the photon emission time on the left-hand side is, so for a second photon emitted at a later time $t_e + \Delta t_e$ we get the same result, and thus

$$\int_{t_e+\Delta t_e}^{t_r+\Delta t_r} \frac{dt}{a(t)} = \int_{t_e}^{t_r} \frac{dt}{a(t)}$$

(this equation essentially defines Δt_r in terms of Δt_e). Generically,

$$\int_{t_e+\Delta t_e}^{t_r+\Delta t_r} (\cdot)\, dt = \int_{t_e}^{t_r} (\cdot)\, dt + \int_{t_r}^{t_r+\Delta t_r} (\cdot)\, dt - \int_{t_e}^{t_e+\Delta t_e} (\cdot)\, dt .$$

Applying this to the previous equation,

$$\int_{t_e}^{t_e+\Delta t_e} \frac{dt}{a(t)} = \int_{t_r}^{t_r+\Delta t_r} \frac{dt}{a(t)} .$$

If we associate Δt with, say, the period P of a visible light wave, $\Delta t \sim 10^{-14}$ sec while $a(t)$ changes significantly only on time scales of many Myr, so for present purposes the above integrands are constants and thus

$$\frac{\Delta t_e}{a(t_e)} = \frac{\Delta t_r}{a(t_r)} \quad \Rightarrow \quad \frac{\Delta t_r}{\Delta t_e} = \frac{a(t_r)}{a(t_e)} . \tag{9.26}$$

Since wavelength λ is proportional to period Δt, this last equation is $\lambda_r/\lambda_e = a(t_r)/a(t_e)$. Thus, from Equation (9.25) the redshift of the received light is

$$z(t_e, t_r) = \frac{a(t_r)}{a(t_e)} - 1 . \tag{9.27}$$

The redshift of light from cosmological objects thus maps out the history of the Universe's expansion between emission and reception times, but in an incomplete manner: only the expansion endpoints are reflected in the redshift. Thus, a set of redshifts spanning the history of the Universe's expansion is required to fully explicate that expansion.

In practice we are most often interested in the redshift observed at the current time, so $t_r = t_0$ and $a(t_r) = 1$ in the above equation. The observed redshift is then

Observed Redshift

$$z(t_0) = \frac{1}{a(t_e)} - 1\,. \tag{9.28}$$

In effect, a galaxy's redshift serves as an observable proxy for its unobservable expansion factor. It is common practice to suppress the explicit time dependencies in this relation and to simply write

$$z + 1 = \frac{1}{a} \tag{9.29}$$

(as we shall frequently do in what follows), with the understanding that z is the redshift observed at the current time and a is for the time of emission of currently observed photons. It is worth noting that the surface of constant redshift is a time slice of the Universe at cosmic time t_e corresponding to Equation (9.28).

Note that nowhere in this derivation has *velocity* been mentioned. Velocity in an expanding Universe is a tricky concept at best and is quite extraneous to the definition of redshift in co-moving coordinates. The units most commonly used in specification of the Hubble Constant – km/sec/Mpc, implying velocities proportional to distance (Equation I.2) – are leftovers from an earlier time when the nature of universal expansion was not properly formulated in relativistic terms, and it would be best for all concerned if such terminology were to be avoided. But specifications such as '$H_0 = 72$ km/sec/Mpc' are so ingrained in astronomers' minds that the mixed units implying recession velocities are the most commonly employed in specifying the Hubble constant.

9.2.2 Time

The cosmic time associated with a value of the expansion function is given by integrating dt/da as given by Equation (9.21):

$$t_a(a) = \int_0^a \frac{dt}{d\tilde{a}} d\tilde{a} = \int_0^a \frac{d\tilde{a}}{d\tilde{a}/dt}\,, \tag{9.30}$$

$$= \frac{1}{H_0} \int_0^a \left[\frac{\Omega_{r,0}}{\tilde{a}^2} + \frac{\Omega_{m,0}}{\tilde{a}} + \tilde{a}^2 \Omega_{\Lambda,0} - (\Omega_0 - 1) \right]^{-1/2} d\tilde{a}\,. \tag{9.31}$$

The current time is

$$t_0 \equiv t_a(1) = \frac{1}{H_0}\int_0^1 \left[\frac{\Omega_{r,0}}{a^2} + \frac{\Omega_{m,0}}{a} + a^2\Omega_{\Lambda,0} - (\Omega_0 - 1)\right]^{-1/2} da\,. \tag{9.32}$$

The cosmic time of emission of photons from a galaxy of redshift z is

$$t_e(z) = \frac{1}{H_0}\int_0^{(z+1)^{-1}} \left[\frac{\Omega_{r,0}}{a^2} + \frac{\Omega_{m,0}}{a} + a^2\Omega_{\Lambda,0} - (\Omega_0 - 1)\right]^{-1/2} da\,. \tag{9.33}$$

The corresponding **look-back time** (the time elapsed since currently observed photons were emitted from a galaxy of redshift z) is

$$\begin{aligned} t_{lb}(z) &\equiv t_0 - t_e(z)\ , \\ &= \frac{1}{H_0}\int_{(z+1)^{-1}}^1 \left[\frac{\Omega_{r,0}}{a^2} + \frac{\Omega_{m,0}}{a} + a^2\Omega_{\Lambda,0} - (\Omega_0 - 1)\right]^{-1/2} da\,. \end{aligned} \tag{9.34}$$

For our flat, matter-only model (Equations (9.23) and (9.24)), $t = t_0 a^{3/2} = t_0(z+1)^{-3/2}$ and $t_0 = 2/(3H_0)$, so these time relations are

$$t_0 = t_a(1) = \frac{2}{3H_0} \qquad \text{(flat, matter only)}, \tag{9.35}$$

$$t_e(z) = t_0\ (z+1)^{-3/2} \qquad \text{(flat, matter only)}, \tag{9.36}$$

$$t_{lb}(z) = t_0\ \left[1 - (z+1)^{-3/2}\right] \qquad \text{(flat, matter only)}. \tag{9.37}$$

These functions of redshift are illustrated in Figure 9.1; similar relations apply to other models. Because of the non-linear relation between observable redshift and unobservable time, the emission and look-back times become rather insensitive to redshift for large-redshift galaxies. From Equation (9.36),

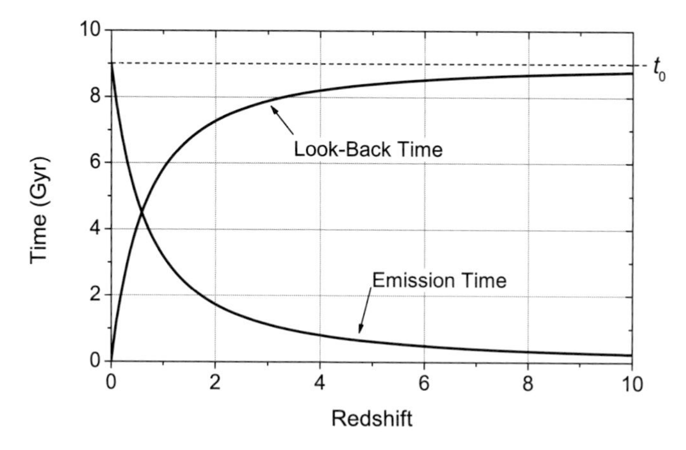

Figure 9.1 Look-back and photon emission times for the flat, matter-only example model with $H_0 = 72$ km/sec/Mpc. The dashed line is the current age (9.009 Gyr) in this model.

$$\frac{\Delta\,(t_e/t_0)}{\Delta z} \approx -\frac{3}{2}\,(z+1)^{-5/2} \qquad \text{(flat, matter only)}.$$

Thus, to advance the observational state of the art from its current $z \approx 5$ to $z = 6$ pushes back the time horizon by less than $\sim$ 2% of the age of the Universe, and at the cost of more than half of the surface brightnesses of observed galaxies (see Problem 8 of this chapter). It is a non-trivial task to observe galaxies earlier and earlier in the Universe's history.

For nearby galaxies ($z \ll 1$) the range of integration for the look-back time is so small that $\dot{a} \approx H_0$ and the look-back time integral (9.34) is approximately $\int_{(z+1)^{-1}}^{1} da = 1 - (\,z+1)^{-1} \approx z$, so that

$$\lim_{z\to 0} t_{\mathrm{lb}}\,(z) = \frac{z}{H_0}\,, \tag{9.38}$$

or $\approx$ 13.5 z Gyr for the canonical $H_0 = 0.074$ Gyr^{-1}. This holds for all cosmological models.

Conformal time

We note that the definition of time used in the RW metric is not unique: any other time variable defined so that g_{00} is a constant and the metric remains diagonal will suffice in principle. In practice it is preferable that the time variable increase with universal expansion, but there are an infinite number of ways to define time so as to meet these requirements: the choice made in the RW metric is merely the simplest possible that also reduces to the local definition of time as kept on ordinary clocks. An alternative choice that offers some advantages is **conformal time**, η, defined by

$$d\eta = \frac{dt}{a} \qquad \Rightarrow \qquad \eta = \int \frac{dt}{a}\,. \tag{9.39}$$

Changing from cosmic to conformal time has the principal effect of expanding time intervals and thus reducing apparent velocities: $\Delta\eta \to \Delta t/\,\langle a \rangle$. Since neither cosmic nor conformal time is the 'correct' system, velocities computed as ratios of distance to time intervals are inherently ambiguous, dependent as they are on the choice of time coordinate. This matter is discussed further in Section 10.4.3.

Cosmological time dilation

Enlargement of time intervals can also arise directly from universal expansion alone. Note from Equation (9.26) that observed time intervals systematically change with redshift according to $\Delta t_0 = (1+z)\,\Delta t_e$. This is a **cosmological time dilation**: events that span a time interval Δt_e in a distant galaxy are observed by us to occupy that time interval multiplied by $1/a = z + 1$; i.e., cosmologically distant clocks appear to run slowly by this factor. This time dilation applies to *all* events, not just EM wave periods; so that, e.g., supernova light curve decay times are observed to be systematically longer in galaxies of larger redshift, by a factor of $z + 1$. Incidentally, this observation constitutes a nice little semi-proof that redshifts really do arise from universal expansion and not from some other physical means, such as intergalactic absorption or 'tired light' (as has been suggested from time to time).

9.2.3 Distance

Cosmologically useful distance measures are based on local measures that conform to the SR definition of distance: we send a photon between the two points in question and infer distance in terms of light travel time: $\Delta d = c\Delta t$. Since light travels along null geodesics, the RW metric for a purely radial path ($d\theta = d\phi = 0$) implies

$$\Delta d = c\Delta t = a(t)\,\frac{\Delta r}{\sqrt{1 - K_0 r^2}}\,, \tag{9.40}$$

in the local limit as $\Delta t, \Delta r \to 0$.

Proper distance (I)

To extend this result to cosmological distances we can proceed in either of two equivalent ways. Operationally, we can imagine a string of closely spaced observers (i.e., galaxies), each of which emits a photon toward the next galaxy in line at the same time t, which effectively serves to define distance as the sum of terms such as that in Equation (9.40). Alternatively, we can imagine the Universe's expansion stopped at time t and use a very long tape measure to determine a galaxy's distance from us at that time, which entails integrating Equation (9.40) along the fixed-time surface labelled x^0 in Figure 8.1. For either procedure the result is the distance defined in our reference frame at a specific time, and is therefore a *proper* distance (as that term is used in SR). For a galaxy of co-moving radial coordinate r_g at time t this **proper distance** d_p is thus defined as

$$d_p(r_g, t) \equiv \left(\int ds\right)_{dt=0} = a(t)\int_0^{r_g} \frac{dr}{\sqrt{1 - K_0 r^2}}\,, \tag{9.41}$$

where we have chosen a coordinate system based on our location and are integrating along the path $d\theta = d\phi = 0$. We know from the redshift discussion that, since $ds = 0$ for a photon travelling from a galaxy to ours,

$$c\int_{t_e}^{t} \frac{dt}{a(t)} = \int_0^{r_g} \frac{dr}{\sqrt{1 - K_0 r^2}}\,,$$

where t_e is the time of photon emission from the galaxy. Comparing these two, the proper distance as a function of photon emission time is

$$d_p(t_e, t) = ca(t) \int_{t_e}^{t} \frac{d\tau}{a(\tau)} . \tag{9.42}$$

The most useful form of proper distance is for the current time and as a function of redshift, denoted by $d_0(z)$. The current proper distance as a function of emission time is given by Equation (9.42) with $t = t_0$ corresponding to $a(t) = 1$. To express this as a function of redshift we note that $dt = da/\dot{a}$ and that $a = (z+1)^{-1}$, so that

$$d_0(z) = c \int_{(z+1)^{-1}}^{1} \frac{1}{a} \frac{da}{da/dt} . \tag{9.43}$$

Now invoking the parameterized Expansion Equation (9.21) for da/dt yields a computationally accessible form for the current proper distance to a galaxy of redshift z.

Current proper distance

$$d_0(z) = \frac{c}{H_0} \int_{(z+1)^{-1}}^{1} \frac{da}{\sqrt{\Omega_{r,0} + a\,\Omega_{m,0} + a^4\,\Omega_{\Lambda,0} - a^2\,(\Omega_0 - 1)}} \tag{9.44}$$

(with $\Omega_0 = \Omega_{r,0} + \Omega_{m,0} + \Omega_{\Lambda,0}$). For nearby galaxies, for which $a \approx 1$ and $z \ll 1$, the integrand in Equation (9.44) is nearly constant and thus

$$d_0(z) \approx \frac{c}{H_0} a \Big|_{(z+1)^{-1}}^{1} = \frac{c}{H_0}\left[1 - (z+1)^{-1}\right] \quad (\text{for } z \ll 1) .$$

It follows that

$$\lim_{z \to 0} d_0(z) = \frac{c}{H_0} z . \tag{9.45}$$

This holds for all cosmological models.

For the flat, matter-only model,

$$d_0(z) = \frac{2c}{H_0}\left[1 - (z+1)^{-1/2}\right] \quad (\text{flat, matter only}). \tag{9.46}$$

This function is illustrated in Figure 9.2. In all expanding models, proper distance is an increasing function of redshift with a negative second derivative, so that proper distance increases with redshift very slowly at large redshifts. In general, proper distances for given redshift are larger in more open geometries.

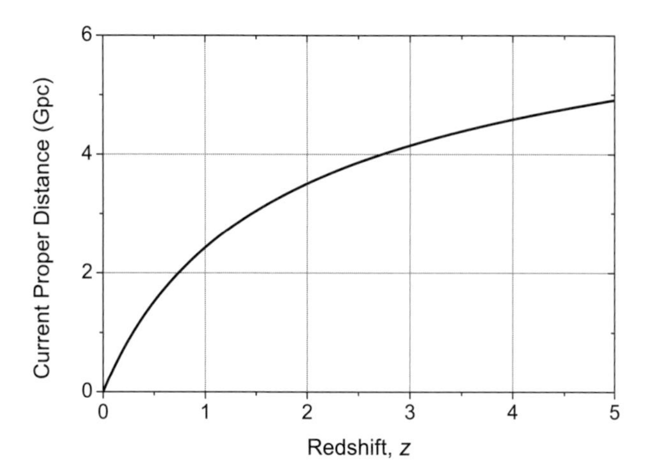

Figure 9.2 Variation of (current) proper distance with redshift for the flat, matter-only model with $H_0 = 72$ km/sec/Mpc.

Proper Distance (II)

The proper distance to a galaxy is not directly observable – it is model-dependent and can only be inferred from observed redshift as in Equation (9.44) – but it forms the basis for connections between cosmological models and definitions of observable distances, and of geometric measures based on distance. The appropriate form of proper distance for this purpose follows from performing the integration in Equation (9.41) :

$$d_{\mathrm{p}}\left(r_{\mathrm{g}}, t\right) = a(t) \times \begin{cases} \arcsin\left(\sqrt{K_0} r_{\mathrm{g}}\right) / \sqrt{K_0}\ , & K_0 > 0, \\ r_{\mathrm{g}}\ , & K_0 = 0, \\ \operatorname{arcsinh}\left(\sqrt{-K_0} r_{\mathrm{g}}\right) / \sqrt{-K_0}\ , & K_0 < 0. \end{cases} \tag{9.47}$$

Three-part function definitions such as this one are going to occur frequently in what follows, and it is worth the effort to formally telescope them by employing the mathematical identities $\arcsin(ix) = i \operatorname{arcsinh}(x)$ and $\sin(ix) = i \sinh(x)$, so that the above equation may be written compactly as

$$d_{\mathrm{p}}\left(r_{\mathrm{g}}, t\right) = \frac{a(t)}{\sqrt{K_0}} \arcsin\left(\sqrt{K_0} r_{\mathrm{g}}\right)\ , \tag{9.48}$$

which holds for all three forms of curvature (for flat geometries, in the limit as $K_0 \to 0$). It further simplifies the notation to modify the definition of radius of curvature so that it is imaginary for negative curvatures and $K = 1/R^2$ in all cases:

$$R_0 \equiv \frac{1}{\sqrt{K_0}} = \frac{c}{H_0} \frac{1}{\sqrt{\Omega_0 - 1}}\ . \tag{9.49}$$

Then the proper distance is given compactly by

$$d_{\rm p}\left(r_{\rm g},t\right)=a\left(t\right)R_0\arcsin\left(\frac{r_{\rm g}}{R_0}\right)\ , \tag{9.50}$$

which is valid for all forms of curvature: in flat geometries, in the limit as $R_0 \to \infty$ where $d_{\rm p}\left(r_{\rm g},t\right)=a\left(t\right)r_{\rm g}$; in open geometries, $\Omega_0 < 1 \Rightarrow R_0$ is imaginary but $d_{\rm p}$ is real:

$$d_{\rm p}\left(r_{\rm g},t\right)=a\left(t\right)\left|R_0\right|\operatorname{arcsinh}\left(\frac{r_{\rm g}}{\left|R_0\right|}\right)\qquad\text{(negative curvature).} \tag{9.51}$$

Note once again that curvature is important only on spatial scales approaching or exceeding $|R_0|$.

The evolution of proper distances of co-moving galaxies follows from the form of Equations (9.48) or (9.50), in which time and space are separable and $d_{\rm p}\left(r_{\rm g},t\right)/a\left(t\right)$ is a function only of the galaxy's co-moving radial coordinate. Then proper distances of co-moving objects at times other than the current are given by

$$d_{\rm p}\left(\cdot,t\right)=a\left(t\right)\,d_0\left(\cdot\right)\ , \tag{9.52}$$

where $(\cdot)$ is a placeholder for the co-moving variable of interest characterizing the galaxy in question. The rate of change of proper distance to a fixed point in the co-moving system (e.g., a galaxy) is then $\dot{d}_{\rm p}=\dot{a}d_0$, or

$$\dot{d}_{\rm p}=Hd_{\rm p}\ . \tag{9.53}$$

This is just Hubble's Law (Equation (I.2)) which thus holds exactly for proper distances.

Related distances

Co-moving distance The **co-moving distance** to a galaxy is one that is tied to the galaxy's radial coordinate and does not change with time. From Equation (9.50) this would be

$$d_{\rm cm}\left(r_{\rm g}\right)=\frac{d_{\rm p}\left(t,r_{\rm g}\right)}{a\left(t\right)}=R_0\arcsin\left(\frac{r_{\rm g}}{R_0}\right)\ , \tag{9.54}$$

where $\arcsin\left(ix\right)=i\operatorname{arcsinh}(x)$ and R_0 is given by Equation (9.49). From Equation (9.52) the co-moving distance of a galaxy of current redshift z is just its current proper distance,

$$d_{\rm cm}\left(z\right)=d_0\left(z\right)\ . \tag{9.55}$$

Radial coordinate The **co-moving radial coordinate** of a galaxy comes from inverting Equation (9.50):

$$r_{\mathrm{g}} = R_0 \sin\left(\frac{d_{\mathrm{p}}}{aR_0}\right) , \tag{9.56}$$

where again $\sin(ix) = i \sinh(x)$ and R_0 is given by Equation (9.49). As a function of redshift as observed at the current time,

$$r_{\mathrm{g}}(z) = R_0 \sin\left(\frac{d_0(z)}{R_0}\right) . \tag{9.57}$$

In a flat geometry this reduces to $r_{\mathrm{g}} = d_0 = d_{\mathrm{cm}}$ by the usual limiting argument. Note that, as with co-moving distance, a galaxy's radial coordinate is an invariant characterization that does not change as the Universe expands.

9.2.4 Areas and volumes

Associated with distance measures are those of surface area and volume. The proper surface area of a sphere centered at the origin, whose surface has co-moving radial coordinate r_{s}, is given by integrating the RW angular metric components over all surface values:

$$A(r_{\mathrm{s}}, t) = \iint [a(t)\, r_{\mathrm{s}}\, d\theta]\, [a(t)\, r_{\mathrm{s}} \sin\theta\, d\phi] = 4\pi a(t)^2 r_{\mathrm{s}}^2 .$$

Invoking Equation (9.56) for r_{s} as a function of proper distance from the center to the surface of the sphere – i.e., the sphere's proper radius – the surface area of a sphere of proper radius d_{p} at time t is then

$$A(d_{\mathrm{p}}, t) = 4\pi \left[a(t)\, R_0 \sin\left(\frac{d_{\mathrm{p}}}{a(t)\, R_0}\right)\right]^2 , \tag{9.58}$$

which reduces trivially to the Euclidean $4\pi d_{\mathrm{p}}^2$ in flat geometries. For a sphere centered on our location, whose surface corresponds to redshift z, the surface area at the current time is

$$A_0(z) = 4\pi \left[R_0 \sin\left(\frac{d_0(z)}{R_0}\right)\right]^2 , \tag{9.59}$$

with the current proper radius d_0 given – in terms of the chosen model – by Equation (9.44), and R_0 by Equation (9.49). Since d_0 is the same as co-moving distance, this last equation describes the co-moving surface area of the sphere of current surface redshift z.

The volume of the sphere is similarly computed as

$$V(r_s, t) = \iiint \left[a(t) \frac{dr}{\sqrt{1 - K_0 r^2}} \right] dA = 4\pi a(t)^3 \int_0^{r_s} \frac{r^2\, dr}{\sqrt{1 - K_0 r^2}}$$

$$\Rightarrow V_0(z) = 2\pi R_0^2 \left[d_0(z) - \frac{R_0}{2} \sin\left(2\frac{d_0(z)}{R_0}\right)\right]. \tag{9.60}$$

This is the co-moving and current proper volume of a sphere centered on the observer and whose surface currently lies at redshift z.

Spherical areas and volumes for given radius are very sensitive to the geometry of space. Since $\sinh x > x > \sin x$, spheres of given radius in open spaces have larger surface areas and volumes than do similar spheres in flat or closed geometries.

9.3 Expansion diagnostics

We choose our cosmological model – i.e., the parameter set $\left(H_0, \Omega_{r,0}, \Omega_{m,0}, \Omega_{\Lambda,0}\right)$, and possibly others – from observations of the real Universe. To some extent this is accomplished by direct observations of energy densities; e.g., $\varepsilon_{m,0}$ from observations of the gravitational effects of matter (galaxy rotations, microlensing, etc.). But this is only part of the story: ultimately, the credibility of a given model lies in its ability to successfully predict observations based on the form of the Expansion Function $a(t)$. That this is so was demonstrated dramatically by the discovery of the energy density represented by the Cosmological Constant, the initial evidence for which was entirely in the form of an observed $a(t)$ that differed from that predicted by our then-standard model, but for which there was then no direct evidence in the form of an observable energy density.

Less dramatically, the form of the Expansion Function reveals itself in variations of observed properties of the Universe's contents. What cosmologists need in order to study $a(t)$ in detail are relations between properties of the Universe's observable contents and the energy densities that modulate the expansion. Particularly useful in this regard are cosmological standards whose inherent physical properties – luminosity, size, etc. – can be relied upon as standard measures. There are not many such, the most useful of which are 'standard candles': objects of known (or reliably presumed) luminosity. The variation of apparent brightness of these standards with redshift is strongly reflective of the expansion history of the Universe during the light's transit time to us, and thus form the basis of the 'brightness–redshift' relations derived from cosmological models. Such theoretical relations are successors to Hubble's original proposal that galaxies were receding from us (inferred from their redshift) at a rate proportional to their distance (inferred from their brightness), and thus are generically known as 'Hubble Relations'. Hubble's work implicitly assumed a flat Universe; more modern forms of the Hubble Relation incorporate possible curvature and are a powerful tool for exploring the Universe's expansion.

9.3.1 Hubble Relation

The variation with distance of brightness of an isotropic emitter is $F = L/A(d)$, where L is the emitter's luminosity (in, e.g., Watts), F is the observed flux (Watts per unit area) at distance d from the emitter, and $A(d)$ is the spherical surface area over which the emitter's isotropic luminosity is evenly spread. In a flat, static Universe $A = 4\pi d^2$; but in curved space-time we must use the relativistic expression of Equations (9.58) or (9.59) for surface area A, and in an expanding Universe we must correct the flux for two relativistic effects: the cosmological redshift which decreases photon energies by a factor of $1/a = z + 1$, and the cosmological time dilation which decreases the photon arrival rate by the same factor (Equation (9.26)); so $F \to F_{\text{static}}/(z+1)^2$ due to expansion. We can neatly incorporate all these matters into the flux computation by defining the **luminosity distance** as

$$\text{luminosity distance:} \qquad d_L(z) \equiv (z+1)\,R_0 \sin\left(\frac{d_0(z)}{R_0}\right). \tag{9.61}$$

The observed flux in an expanding, (possibly) curved Universe is then given by

$$F(z) = \frac{L}{4\pi\, d_L(z)^2}. \tag{9.62}$$

The cosmographical significance of these results is that, unlike the case with proper distance, d_L is effectively an observable distance to a galaxy:

$$d_L = \left(\frac{L}{4\pi F}\right)^{1/2}. \tag{9.63}$$

We 'observe' d_L for a galaxy by directly measuring its flux F and presuming its luminosity L.

Combining Equations (9.61) and (9.62):

$$F(z) = \frac{1}{4\pi}\frac{L}{(z+1)^2\left[R_0 \sin\left(d_0(z)/R_0\right)\right]^2}. \tag{9.64}$$

This is the basic form of the **Hubble Relation**, connecting observed brightness to redshift in a model-dependent manner. In practice, astronomers usually prefer to deal with apparent (m) and absolute (M) magnitudes rather than fluxes and luminosities; the two systems are related by $m = -2.5\log F + C_1$ and $M = -2.5\log L + C_2$ for some constants C defining the photometric system being used. Apparent (observed) and absolute (assumed) magnitudes are then related to distance by

$$m(z) - M = 5 \log d_L(z) + \mu_0 , \tag{9.65}$$

where μ_0 is a calibrating distance modulus: $\mu_0 = (25, 40)$ for d_L in units of (Mpc, Gpc), respectively. The quantity $\mu \equiv m - M$ is commonly known as the **distance modulus**.

The Hubble Relation can then be employed as one relating distance to redshift (as revealed in traditional Hubble diagrams), or as one directly relating observed brightness to redshift: from Equations (9.61) and (9.65),

Hubble Relations

$$d_L(z) \equiv \left(\frac{L}{4\pi F}\right)^{1/2} = (z+1) R_0 \sin\left[\frac{d_0(z)}{R_0}\right] , \tag{9.66}$$

$$\mu(z) \equiv m(z) - M = 5 \log\left[(z+1) R_0 \sin\left(\frac{d_0(z)}{R_0}\right)\right] + \mu_0, \tag{9.67}$$

where $d_0(z)$ and R_0 are computed for the chosen model by Equations (9.44) and (9.49), respectively; and $\mu_0 = (25, 40)$ for distances expressed in units of (Mpc, Gpc), respectively. One observes (F, z) or (m, z) for standard candles and compares them to the relations implied by the above equations in order to assess the credibility of the presumed cosmological model. Examples of these two relations for our flat, matter-only example model are shown in Figure 9.3. Note that for small redshifts, $d_L(z) \approx (z+1) d_0(z)$; from Equation (9.45) then, the Hubble Relations for nearby galaxies obey

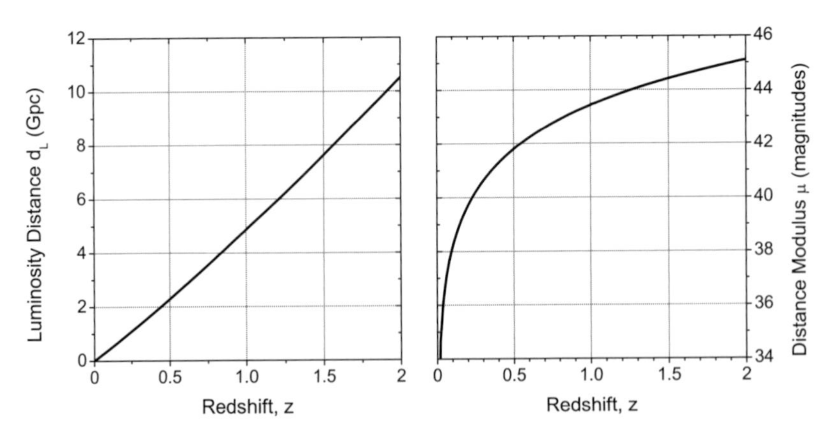

Figure 9.3 Hubble Relations (Equations (9.66) and (9.67), respectively) for the flat, matter-only (Einstein–de Sitter) cosmological model ($H_0 = 72$ km/sec/Mpc).

$$\lim_{z\to 0} d_{\mathrm{L}} = \frac{c}{H_0} z \,, \tag{9.68}$$

$$\lim_{z\to 0} \mu = 5 \log \left(\frac{c}{H_0} z \right) + \mu_0 \,. \tag{9.69}$$

These hold for all cosmological models. As claimed in Chapter 3, the Hubble Constant can reliably be estimated from the slope of the distance–redshift relation at small redshifts (if distances are measured as luminosity distances, usually via Equation (9.66)).

The attentive observational astronomer will have noticed that many details were left out of the photometry discussion here. In particular, photometric quantities need to be confined to finite passbands if they are to be observable, a practical necessity that requires 'K-corrections' to observed brightnesses in order to track the change of passbands with redshift; and observed fluxes and magnitudes must be corrected for such extraneous things as atmospheric absorption and instrumental efficiency. Consult any good book on astronomical photometry for the many practical details inherent in brightness measurement.

9.3.2 Angular diameters

In a flat, static Universe the apparent angular size of a galaxy of diameter D would be

$$\Delta\theta = \frac{D}{d} \quad \text{(flat, static)} \,, \tag{9.70}$$

where d is the distance to the galaxy, presumed $\gg D$. But in an expanding or curved Universe the dependence of angular diameter on distance is more complicated.

To appreciate the effects of curvature, consider measurement of angular diameters conducted on the two-dimensional surface of a sphere. With the origin at one pole, the angular diameter of an object is the angle subtended by the two meridians encompassing it. This angle decreases as the object recedes to the 'equator', but then begins to increase again as the object moves toward the other pole, a purely geometric effect arising from the curvature of the surface; so that distant objects can appear larger than closer ones.

A related phenomenon occurs even in flat geometries if the space is expanding: we see a distant object not as it is now, but as it was some time ago when it was closer to us. In an isotropic and uniformly expanding Universe (i.e., expanding the same in all directions) the angle subtended by two light rays does not change with expansion, so that the angular diameter of an object is always that at the time when the light was emitted. Ongoing expansion thus leads to larger apparent angular diameters of distant objects than would be seen in a static Universe.

The upshot of both effects is that the angular diameters of galaxies typically appear to decrease with redshift up to a certain point, then stop decreasing and begin increasing after that in a manner reflective of the curvature and expansion history of the Universe. The correct expression for angular diameter as a function of redshift comes, once again, from analysis with the Robertson–Walker metric. With both time and radial coordinates fixed, and angular measures confined to the θ-direction, the metric across the face of a distant galaxy is

$$ds^2 = a\left(t_e\right)^2 r_g^2 \, d\theta^2 \, .$$

Taking the square root and integrating across the face of the galaxy, $\int ds = D$ (the diameter) and $\int d\theta = \Delta\theta$ (the angular diameter); so the apparent angular diameter is $\Delta\theta = D \big/ \left[a\left(t_e\right) r_g\right]$. Then from Equation (9.57) for $r_g(z)$,

$$\Delta\theta\left(z\right) = \frac{D}{R_0} \frac{\left(z+1\right)}{\sin\left[d_0\left(z\right)/R_0\right]} \, , \tag{9.71}$$

at the current time. Figure 9.4 illustrates this angular diameter vs. redshift relation for the flat, matter-only model.

Similar to what was done with the Hubble Relation, we can define an angular diameter distance as

$$d_A\left(z\right) \equiv \frac{1}{z+1} R_0 \sin\left[\frac{d_0\left(z\right)}{R_0}\right] \, , \tag{9.72}$$

$$= \frac{d_L\left(z\right)}{\left(1+z\right)^2} \quad \text{(from Equation (9.61))} \, ; \tag{9.73}$$

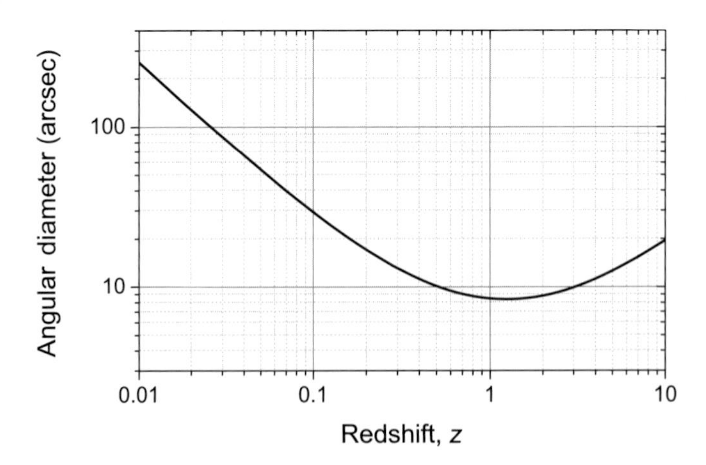

Figure 9.4 Angular diameter variation with redshift of a 50 kpc diameter galaxy in the flat, matter-only model for $H_0 = 72$ km/sec/Mpc. Note logarithmic scaling on both axes.

in terms of which

$$\Delta\theta(z) = \frac{D}{d_A(z)}. \tag{9.74}$$

9.3.3 Galaxy densities

An additional diagnostic of expansion models – historically important, if seldom used now – is one of measurements of the spatial number densities of galaxies. If galaxies are neither created nor destroyed, the number in a co-moving volume will not change, but the spatial volume encompassing them (defined by coordinate values, or by z) will increase in a manner controlled by $a(t)$ and reflective of the geometry of space. Thus, the number *density* of galaxies in the volume out to a limiting redshift will vary with redshift in a manner that reflects the geometry and expansion history of the Universe.[1]

As usual, let the observer be at the origin and consider the volume in a co-moving sphere of radius defined by $z < z_L$ and within the solid angle $\Delta\omega = \sin\theta\, d\theta\, d\phi$; this is the sphere's volume given by Equation (9.60), multiplied by $\Delta\omega/4\pi$ (for small $\Delta\omega$). If galaxies are not being created or destroyed, the number within this volume at the current time will be

$$N(z < z_L) = \frac{n_0}{2}R_0^2\left[d_0(z_L) - \frac{R_0}{2}\sin\left(2\frac{d_0(z_L)}{R_0}\right)\right]\Delta\omega, \tag{9.75}$$

where n_0 is the current local number density of galaxies. This cumulative density can be converted to a differential one by differentiating with respect to z:

$$\mathsf{n}(z) = \frac{d}{dz}N(z) = \frac{n_0}{2}R_0^2 d_0'(z)\left[1 - \cos\left(2\frac{d_0(z_L)}{R_0}\right)\right]\Delta\omega, \tag{9.76}$$

where

$$\begin{aligned} d_0'(z) &\equiv \frac{d}{dz}d_0(z) \\ &= \frac{c}{H_0}\left[(z+1)^4\Omega_{r,0} + (z+1)^3\Omega_{m,0} + \Omega_{\Lambda,0} - (z+1)^2(\Omega_0 - 1)\right]^{-1/2}. \end{aligned} \tag{9.77}$$

The observable quantity $\mathsf{n}(z)\,dz$ is the number of sources with redshifts lying between z and $z + dz$, within solid angle $\Delta\omega$. More generally,

$$N(z < z_2) - N(z < z_1) = \int_{z_1}^{z_2} \mathsf{n}(z)\; dz \tag{9.78}$$

[1] This was originally done with radio 'sources' whose nature was then unknown. We now know that such sources are galaxies, but the method is still most frequently referred to as 'source counts' or 'source densities'.

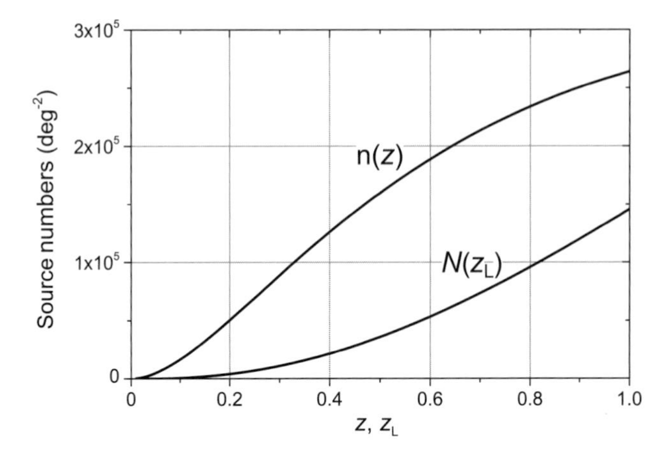

Figure 9.5 Cumulative (N; Equation (9.75)) and differential (n; Equation (9.76)) source counts for $n_0 = 10^8$ Gpc^{-3} in a flat, matter-only model with $H_0 = 72$ km/sec/Mpc.

is the number of sources with $z_1 < z < z_2$, within solid angle $\Delta\omega$. These functions are demonstrated in Figure 9.5.

9.3.4 Deceleration: q_0 and friends

In the early days of modern cosmology, when 'high redshift' galaxies had $z \sim 0.1$, it made sense to approximate the Hubble Relation with a low-order Taylor's series expansion in z. From Equation (9.68) the first term in that expansion would be $d_L \approx (c/H_0)\, z$ and the second term, involving z^2, would reflect the first-order departure of the relation from linearity. To relate this observable feature to the expansion function the non-dimensional **deceleration parameter** q and **deceleration constant** $q_0 \equiv q\,(t_0)$ were defined as

$$q \equiv -\frac{\ddot{a}a}{\dot{a}^2}\,, \tag{9.79}$$

$$\Rightarrow q_0 = -\frac{\ddot{a}_0}{\dot{a}_0^2} = -\frac{\ddot{a}_0}{H_0^2}\,. \tag{9.80}$$

In terms of q_0 the Taylor's series representation of $d_L\,(z)$ is

$$d_L = \frac{c}{H_0}\left[z + \frac{1}{2}\,(1 - q_0)\, z^2 + \cdots\right]. \tag{9.81}$$

Higher order terms and constants can be similarly defined: the next, entitled the *Jerk*, is given by $j \equiv \dddot{a}a^2/\dot{a}^3$. Even higher-order terms are whimsically called *Snap, Crackle, Pop,* etc. (I'm not making this up); see Weinberg (2008) for further discussion of this representation of the Hubble Relation. In the current era, with observable redshifts $\gtrsim 1$, this power series approach is not very useful and most cosmologists prefer to use the exact form of the Hubble Relation given by Equations (9.66) and (9.44) (as applied to actual models). But the deceleration constant is ubiquitous in the literature, particularly in older texts, and it is worth the effort to derive its relation to the more modern parameterization involving energy densities.

From the Acceleration Equation (9.22) we have, for the current time:

$$\ddot{a}(t_0) = -H_0^2\left[\Omega_{r,0} + \frac{1}{2}\Omega_{m,0} - \Omega_{\Lambda,0}\right] ;$$

and from the definition of q_0 we have $\ddot{a}_0 = -H_0^2 q_0$. Equating these,

$$q_0 = \Omega_{r,0} + \frac{1}{2}\,\Omega_{m,0} - \Omega_{\Lambda,0} \,. \tag{9.82}$$

It is apparent that q_0 is an incomplete proxy for the energy density parameters. In the current cosmological era when $\Omega_{r,0}$ is insignificant, and prior to $\sim$ 1998 when the presence of the cosmological constant was revealed, one could credibly write $q_0 = \frac{1}{2}\,\Omega_0$ and use either parameter in the Friedmann Equations. In particular, q_0 as inferred from the Taylor's series expansion for the observed Hubble Relation could be used to estimate curvature: the Universe is (closed, flat, open) if q_0 is $\left(> \frac{1}{2},\ = \frac{1}{2},\ < \frac{1}{2}\right)$, respectively. But more modern treatments tend to favor the use of energy density parameters over q_0 since, among other things, they contain more information.

Problems

1. For a flat model containing only radiation, and currently expanding at rate $H_0 = 0.074\ \text{Gyr}^{-1}$:
   (a) show that the expansion function obeys $a(t) \propto t^{1/2}$ and find the constant of proportionality;
   (b) numerically evaluate the current radiation energy density in units of J/m^3;
   (c) what is the current age of this model, in units of Gyr?
2. For a model Universe with $a(t) \propto t^{3/5}$ and Hubble constant H_0,
   (a) find an expression for the current proper distance d_0 in terms of H_0 and the redshift z;
   (b) if the model is also flat, find an expression for the luminosity distance in terms of z;
   (c) if this flat model Universe contains only one gravitating component, compute its EOS parameter w (assumed to be constant).

3. Consider a flat Universe whose gravitating content has an equation of state parameter $w = 2/3$.

   (a) Set up and solve the first Friedmann Equation for the function $a(t)$.
   (b) If the current energy density of the Universe is $\varepsilon_0 = 8.8 \times 10^{-10}$ J/m^3, find values for the Hubble Constant H_0 (in units of Gyr^{-1}) and for the current age of the Universe, t_0 (in Gyr).
   (c) Find an expression for the Hubble Relation, $d_L(z)$. Show that the slope of this relation near the origin is c/H_0.

4. Find an expression for the Hubble relation in a flat Universe that contains only one gravitating component with a constant EOS parameter $w \neq -1/3, -1$. Then find the flux observed from a standard candle of luminosity L in this model if its redshift is z, and show that the flux is a monotonically increasing function of w. Comment on observational consequences of the EOS parameter.
5. A photon is sent from our galaxy to another located 1 Gpc away (current proper distance, d_0). In the flat, matter-only model with $H_0 = 72$ km/sec/Mpc, how long does it take for the photon to arrive, and what is the galaxy's proper distance at the arrival time?
6. For the galaxy and model of the previous problem, what are the galaxy's redshifts at the current time and at the time of photon arrival there?
7. For an empty Universe, $\Omega_{x,0} = 0$ for all components x. Show that such a model expands at a constant rate and find expressions for K_0 and R_0 for such a model, in terms of H_0. Use these to find expressions for proper and luminosity distances in terms of redshifts.
8. The surface brightness of a galaxy is the flux per unit solid angle as seen at the telescope:

$$B_{\text{surface}} = \frac{F}{\pi\,(\Delta\theta/2)^2} \approx \frac{F}{\Delta\theta^2},$$

   where F is the observed flux and the galaxy has apparent angular diameter $\Delta\theta$ (in steradians). Use the definitions of luminosity distance and angular diameter distance to show that, for a galaxy of given diameter D and luminosity L, its surface brightness is a function of redshift that is independent of the cosmological model, and find the form of that relation. For a fixed D and L, by what factor does the surface brightness differ from a galaxy of redshift $z = 1$ to one with $z = 3$? Since the signal detected on a CCD detector is proportional to the surface brightness, what are the implications for observational cosmology?
9. Show that for flat geometries, Equations (9.61), (9.60), (9.71), and (9.75) reduce to, respectively:

$$d_L(z) = (z+1)\, d_0(z) \ , \tag{9.83}$$

$$V(z) = \frac{4}{3}\pi \left[d_0(z)\right]^3 \ , \tag{9.84}$$

$$\Delta\theta(z) = (z+1)\frac{D}{d_0(z)} \ , \tag{9.85}$$

$$N(z < z_L) = \frac{1}{3} n(t_0) \left[d_0(z_L)\right]^3 \Delta\omega \ . \tag{9.86}$$

10. Show that the circumference of a circle centered at the coordinate origin and of RW coordinate radius r_c is given by $C = 2\pi r_c$. In terms of proper distance d corresponding to r_c, show that $C(d)$ is $(<, =, >) 2\pi d$ in spaces of (positive, zero, negative) curvatures, respectively; and that $C(d)$ conforms to the Bertrand–Diquet–Puiseux formula (Equation (6.3)) in all three cases.
11. Show that for an expansion model of the form $a(t) \propto t^x, x \neq 1$, the Hubble distance–redshift relation is

$$d_L(z) = \frac{x}{x-1}\frac{c}{H_0}\left[z + 1 - (z+1)^{(2x-1)/x}\right] \ ,$$

in a flat geometry. What is the equivalent expression for the case $x = 1$?
12. The inference of universal expansion arises from the observation that galaxy redshifts increase with distance. But this could also be a consequence of energy loss by photons on their way to us. This is the 'tired light' hypothesis, the simplest form of which is a static Universe in which photon energies decrease with distance r in a manner similar to that of interstellar absorption: $\Delta\varepsilon = -\kappa\varepsilon\Delta r$ for some positive constant κ ('absorption coefficient'). Investigate the cosmography of such a model, and show that to some extent it can mimic that of expanding models. In particular, derive expressions for z as a function of r; the Hubble Relation $d_L(z)$; and photon enroute time $t_{LB}(z)$. Use the small redshift approximation to the Hubble Relation to estimate κ for the canonical $H_0 = 72$ km/sec/Mpc.

10

Expansion dynamics

10.1 Curvature

At sufficiently early times the dominant source of gravitating energy was radiation, a consequence of its a^{-4} dependence on expansion; see Figure 8.2. For our current observational estimate of $\Omega_{r,0} \approx 10^{-4}$, radiation energy densities will exceed those of matter and cosmological constant combined when $a \lesssim 10^{-4}$. At such early times the Expansion Equation was

$$\dot{a}^2 \approx H_0^2 \frac{\Omega_{r,0}}{a^2} - K_0 c^2 \quad \text{(radiation only)} . \tag{10.1}$$

The first (energy) term on the right-hand side will dominate the second (curvature) term for sufficiently small values of a, whereupon the Universe's expansion rate will effectively be independent of its curvature. The extent of this dynamical decoupling is non-negligible for much of the early Universe: from the above equation and Equation (9.17) for geometric curvature, the ratio of the curvature term to that of energy is

$$\left| \frac{\text{curvature}}{\text{energy}} \right| \longrightarrow \frac{|K_0|\, c^2}{H_0^2\, \Omega_{r,0}} a^2 = \frac{|\Omega_0 - 1|}{\Omega_{r,0}} a^2 . \tag{10.2}$$

Since $|\Omega_0 - 1| < 1$ in all plausible models, and $\Omega_{r,0}$ is observed to be $\sim 10^{-4}$, this ratio is $\lesssim 10^{-4}$ throughout the radiation-dominated era when $a \lesssim 10^{-4}$. Put succinctly:

> the early Universe was radiation-dominated and expanded *as if it were flat*, irrespective of its geometric curvature.

We thus expect the expansion of the early Universe to obey the simple expansion equation $\dot{a}^2 \approx H_0^2 \Omega_{r,0}/a^2$ up to the point of equality of radiation and matter energy densities which – depending upon the specific model adopted – probably occurred at an age between $\sim 10^4$ and $\sim 10^5$ years. Prior to then the Universe was, for all practical purposes, flat.

A similar situation arises in the late Universe when dark energy dominates the expansion, and the Expansion Equation (9.21) effectively becomes

$$\dot{a}^2 \approx H_0^2 \left[a^2 \Omega_{\Lambda,0} - (\Omega_0 - 1)\right] \quad (a \gg 1) \ . \tag{10.3}$$

For sufficiently large a the energy term $a^2\Omega_{\Lambda,0}$ greatly dominates the curvature term $(\Omega_0 - 1)$ and the Universe expands as if it were flat. It is thus only in the intermediate stages of expansion, when a is within a few orders of magnitude of 1, that geometric curvature may play a role in the Universe's expansion; and then only if the current curvature is non-zero.

From Equation (9.17) or (9.18), cosmological curvature is seen to arise from two things: energy densities (Ω) and expansion rate (H). In particular, a non-zero curvature is possible even in an empty Universe, as discussed in Section 8.6; and can therefore (if you wish) be considered a property of space itself. But – as argued in that section – curvature must be considered to be a metric quantity, dependent upon the coordinates chosen in which to express universal expansion. As a consequence, physical properties involving curvature are inherently ambiguous. The matter is discussed further in Section 10.4.

10.2 Expansion fate

The gravitation of matter and radiation act to slow the Universe's expansion, while dark energy acts to accelerate it. For nearly all the Universe's history its expansion has been driven by one or both of matter and Cosmological Constant. The corresponding single-component versions of the Expansion Equation (9.21) are

$$\left(\frac{\dot{a}}{H_0}\right)^2 = \frac{\Omega_{\mathrm{m},0}}{a} - \left(\Omega_{\mathrm{m},0} - 1\right) \qquad \text{(matter only)} \ , \tag{10.4}$$

$$\left(\frac{\dot{a}}{H_0}\right)^2 = \Omega_{\Lambda,0}\, a^2 - \left(\Omega_{\Lambda,0} - 1\right) \qquad (\Lambda \text{ only}) \ . \tag{10.5}$$

The attractive gravitation of matter acts to slow the expansion rate so that, for given current rate H_0, matter-only models with larger $\Omega_{\mathrm{m},0}$ were expanding more rapidly in the past than were models with smaller values of $\Omega_{\mathrm{m},0}$. The Cosmological Constant, on the other hand, represents a force of repulsion and thus acts to accelerate the expansion, so that Λ-only models with higher values of $\Omega_{\Lambda,0}$ were expanding more slowly in the past than were those with smaller values (for given H_0). These effects are illustrated in Figures 12.1 and 13.2 for various values of the two components.

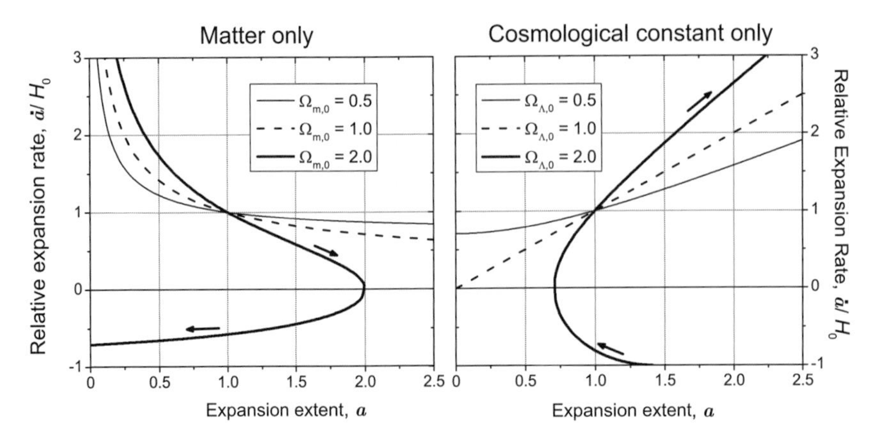

Figure 10.1 Evolution of relative expansion rates, $\dot{a}/H_0$, in single-component models (see text). Arrows denote increasing time.

For $\Omega_0 < 1$ both matter-only and Λ-only models have $\dot{a}^2 > 0$ for all values of a and thus may expand forever. But overly dense models ($\Omega_0 > 1$) are limited in their expansion. For matter only, models with $\Omega_{m,0} > 1$ have their expansion rate go to zero at $a_{max} = \Omega_{m,0}/\left(\Omega_{m,0} - 1\right) > 1$, following which they re-contract. For models with only Cosmological Constant, those with $\Omega_{\Lambda,0} > 1$ cannot begin expansion with $\dot{a} > 0$ without their current expansion rate exceeding H_0, and so have a minimum expansion extent: $a > a_{min} = \left[\left(\Omega_{\Lambda,0} - 1\right) / \Omega_{\Lambda,0}\right]^{1/2}$. These limitations are clearly seen as phase trajectories in Figure 10.1.

Except for the earliest times – e.g., prior to galaxy formation – the Universe's expansion appears to have been governed by a mix of matter and Λ-dark energy and so is a combination of the behaviors exhibited in Figure 10.1. In general, the expansion rates decrease with a in the early stages and increase later, when the Cosmological Constant takes over. Examples demonstrating the corresponding range of possibilities in expansion histories are shown in Figure 10.2.

The heavy solid curve denotes our current best estimate for our Universe, the CCM of Chapter 15. The thinner solid curve $\left(\Omega_{\Lambda,0} = 1.665\right)$ is an example of the 'loitering' models first studied by Georges Lemaître in the 1920s: they are characterized by $\dot{a}$ becoming very nearly equal to 0 at some value of a so that the expansion nearly, but not quite, stops; then resumes at a later time. The unique observable property of such universes is that they concentrate a lot of history near one value of a or z; here, $z \approx 1$ for more than 30 Gyr.

The highest density model – graphed as a dash-dot curve – is an example of an over-dense model whose Λ-dark energy by itself would require an initially contracting phase if the current expansion rate were not to exceed H_0; in concert with matter it restricts the possibilities to (1) an initially expanding model that stops expanding and recontracts at a value of $a < 1$ (lower branch), or (2) begins at $t = 0$ as a contracting universe that stops its contraction at $a < 1$ and

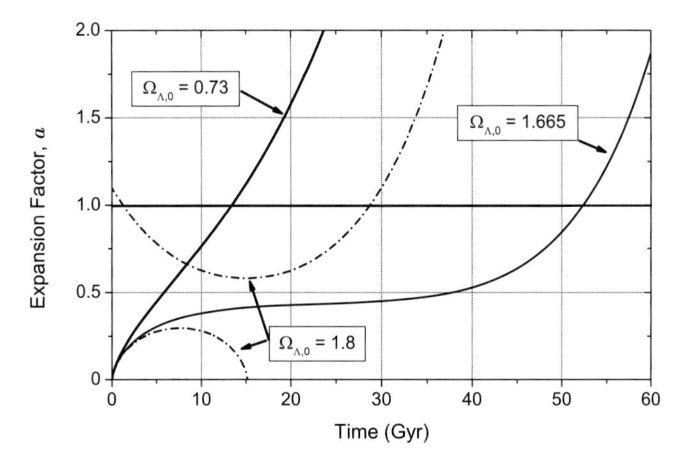

Figure 10.2 Expansion histories for some two-component models, all with $\Omega_{\mathrm{m},0} = 0.27$ and $H_0 = 72$ km/sec/Mpc $= 0.074\ \mathrm{Gyr}^{-1}$ (see text).

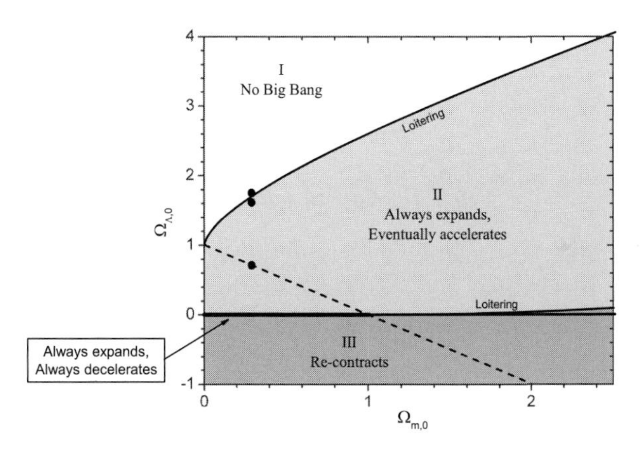

Figure 10.3 Evolutionary fates of universes with different matter and cosmological constant energy densities, all for $H_0 = 0.074\ \mathrm{Gyr}^{-1}$. The dashed line denotes geometrically flat models $\left(\Omega_{\Lambda,0} + \Omega_{\mathrm{m},0} = 1\right)$ and divides regions with positive (above) and negative (below) curvature. The solid circles denote parameters for the models graphed in Figure 10.2; the lowest one is our best estimate for our Universe and corresponds to the heavy solid curve in Figure 10.2.

re-expands (upper branch). In the first case the Universe never expands to its current size, and in the second it never experiences a hot, dense early phase: models with $\Omega_{\Lambda,0}$ greater than some critical value (dependent upon $\Omega_{m,0}$) thus cannot represent our Universe.

The expansion fates of models of differing compositions of matter and Λ-dark energy are summarized graphically in Figure 10.3, where the possibilities are gathered into three regions in the $\Omega_{\Lambda,0}-\Omega_{\mathrm{m},0}$ plane. Models in region II (which almost certainly includes our Universe) initially decelerate as they expand, then accelerate as $\Omega_{\Lambda,0}$ comes to command the expansion and subsequently expand forever at ever increasing rates, much as demonstrated by the solid curves in Figure 10.2. Models in region III stop their expansion at some value of $a > 1$, then re-contract. Models in region I either expand until $a = a_c < 1$ and then re-contract, or must initially contract from $a > 1$ and turn around at some $a < 1$ and re-expand from there, much as shown by the dash-dot curve in Figure 10.2; and thus are not realistic models for our Universe.

The labelled (lower left) heavy horizontal line corresponds to models with $\Omega_{\Lambda,0} = 0$ and $\Omega_{m,0} < 1$: these are the classical 'open' universe models and are the only ones that expand forever without entering an accelerating phase. Note that models with both positive curvature ($K > 0$; above the dashed line) and negative curvature ($K < 0$; below it) can either expand forever or re-contract, depending upon the values of the energy densities. In a universe with non-zero cosmological constant, the terms 'open' and 'closed' are purely geometric and largely disconnected from the eventual fate of the expansion.

10.3 Horizons

10.3.1 Causal connections

Consider an infinite, non-relativistic, static universe of age t_0. The furthest light could have travelled in that time is ct_0, so not only is that the furthest we could see but it also marks the boundary between events that could have influenced our observable Universe ($d < ct_0$) and those that are outside our influence. This is a **horizon**, a quantity that effectively defines the size of the Universe that we can experience; i.e., with which we are in causal contact. Horizons are useful quantities in studying the dynamics of universal expansion, for anything outside the horizon need not be considered in such analyses. In an expanding Universe the situation is, as you might expect, a bit more complicated; and it is useful to expand the concept of 'horizon' to one of causal limits in general.

Signals – of whatever nature – propagate amongst fundamental observers at rates dependent upon both the signal velocity in co-moving coordinates, and the Universe's expansion. If the signal propagates at velocity v with respect to co-moving observers, its proper distance increases at rate $\dot{d}_p$ given by Equation (9.53), augmented by v; so that $\dot{d}_p = Hd_p + v$. Re-arranging, $(a\dot{d}_p - \dot{a}d_p)/a^2 = v/a$, which integrates immediately to

$$d_p(t_e, t_r) = a(t_r) \int_{t_e}^{t_r} \frac{v(\tau)}{a(\tau)} d\tau \,. \tag{10.6}$$

This is the distance a signal of velocity v can travel between times t_e and t_r, and is thus the causal limit for the signal being propagated. In particular, if the signal is one of light this formula just recovers the proper distance between two fundamental observers as given in Section 9.2.3:

$$d_{\mathrm{p}}(t_{\mathrm{e}}, t) = c\, a(t) \int_{t_{\mathrm{e}}}^{t} \frac{d\tau}{a(\tau)} . \tag{10.7}$$

The representation of causal barriers as horizons can usefully be employed in many ways, of which the most common are the following.

10.3.2 Hubble horizon

Applying the identity $dt = da/\dot{a}$ and the definition $H \equiv \dot{a}/a$ to Equation (10.7),

$$d_{\mathrm{p}}(t_{\mathrm{r}}) = c\, a(t_{\mathrm{r}}) \int_{a(t_{\mathrm{e}})}^{a(t_{\mathrm{r}})} \frac{da}{a^2 H} .$$

If we confine the integration to a range over which H is nearly constant,

$$d_{\mathrm{p}} \approx \frac{c\, a(t_{\mathrm{r}})}{H} \left[\frac{1}{a(t_{\mathrm{e}})} - \frac{1}{a(t_{\mathrm{r}})} \right] = \frac{c}{H} \left[\frac{a(t_{\mathrm{r}})}{a(t_{\mathrm{e}})} - 1 \right] .$$

Thus, in the time needed for the Universe to double in size light travels the approximate distance given by the **Hubble horizon**:

$$d_{\mathrm{H}} = \frac{c}{H} . \tag{10.8}$$

This is the same as the Hubble distance defined in Chapter 9 and is a characteristic distance scale in an expanding Universe, often (if rather imprecisely) simply called 'the horizon'. Its current value in the CCM is ≈ 4.15 Gpc.

10.3.3 Particle horizon

The **Particle horizon**[1] is a causality barrier in the past: the distance to the most remote object with which we have been in causal contact; i.e., from which light could have reached and influenced us since the Big Bang. It is normally defined as the proper distance travelled by light since $t = 0$, so from Equation (10.7) the particle horizon at time t is

$$d_{\mathrm{PH}}(t) = c\, a(t) \int_{0}^{t} \frac{d\tau}{a(\tau)} . \tag{10.9}$$

[1] The name *particle horizon* derives from the conceptual foundations of relativistic cosmology, wherein co-moving galaxies are "fundamental particles" making up the expanding Universe. The term was originally proposed by the Austrian physicist Wolfgang Rindler in the mid-twentieth century.

This is a monotonically increasing function of time in any expanding model. Differentiating the above expression with respect to time yields the horizon expansion rate:

$$\dot{d}_{\mathrm{PH}}(t) = H(t)\, d_{\mathrm{PH}}(t) + c\,. \tag{10.10}$$

Horizon growth in an expanding Universe can thus be viewed as a combination of a pure speed-of-light causality (c), plus growth due to expansion of the co-moving coordinate system ($H\, d_{\mathrm{PH}}$). In particular, in a static Universe $H(t) = 0$ and the particle horizon grows as $d_{\mathrm{PH}}(t) = ct$ (static Universe), while in an expanding Universe the horizon grows more rapidly.

In our flat, matter-only model, $a(t) = (t/t_0)^{2/3}$ and the particle horizon is easily seen to be $d_{\mathrm{PH}}(t) = 3ct$ while proper distances go as $d_{\mathrm{p}}(\cdot, t) = a(t)\, d_0(\cdot) \propto t^{2/3}$: the horizon eventually grows more rapidly than do proper distances to galaxies, so more and more galaxies enter our horizon as time goes on. In the de Sitter model, $a(t) \propto \exp(H_0 t)$ and the particle horizon grows as $d_{\mathrm{PH}}(t) = (c/H_0)\left[\exp(H_0 t) - 1\right]$ while proper distances are $d_{\mathrm{p}}(\cdot, t) = a(t)\, d_0(\cdot) \propto \exp(H_0 t)$: eventually the horizon grows at the same relative rate as do proper distances, and the number of galaxies within the horizon remains constant.

It is sometimes useful to compute particle horizons from some initial time other than $t = 0$. In particular, the currently accepted model of universal expansion includes an early ($t \approx 10^{-36}$–10^{-34} sec), exponential expansion of many orders of magnitude ('inflation') that greatly enlarges particle horizons beyond the values computed from $t = 0$. In a useful sense the end of inflation can be regarded as the 'beginning of the Universe' and the particle horizon distances most commonly used are those computed from the time of inflation's end (t_{i0}). From Equation (10.7),

$$d_{\mathrm{PH}}(t_{\mathrm{ie}}, t) = c\, a(t) \int_{t_{\mathrm{ie}}}^{t} \frac{d\tau}{a(\tau)}. \tag{10.11}$$

Inflation (and its effects on horizons) is discussed in detail in Chapter 16.

10.3.4 Event horizon

Event horizons are causality barriers in the future, the distance that light will be able to travel from now until $t = \infty$. Consider a photon emitted at the current time from a galaxy of radial coordinate r_{g} and proper distance d_0. Since $ds = 0$ along the photon's path it will be received here at future time t_{r} given implicitly by the usual integral

$$c \int_{t_0}^{t_{\mathrm{r}}} \frac{dt}{a(t)} = \int_0^{r_{\mathrm{g}}} \frac{dr}{\sqrt{1 - K_0 r^2}} \qquad (= d_0)$$

(see Equation (9.42)). The right-hand side of this equation depends only on the galaxy's co-moving location, while the left-hand side is dependent upon the form of the expansion factor $a(t)$: different such factors (i.e., expansion models) will lead to different light reception times t_r for the same galaxy. For flat models containing only matter and radiation – for which $a(t) \propto t^n$ with $0 < n < 1$ – the left-hand integral above becomes arbitrarily large as $t_r \to \infty$, so the reception time t_r will be finite for any finite current proper distance, no matter how large: all galaxies in the Universe will eventually become visible. But when Λ-dark energy dominates, the future expansion goes as $a(t) \approx \exp[H(t - t_0)]$ and the left-hand integral stays finite as $t_r \longrightarrow \infty$: there is a maximum (current) proper distance from which signals can be emitted now and be received at some time in the future. This limiting proper distance for an infinite photon travel time is the **event horizon**:

$$d_{EH}(t) \equiv c\, a(t) \int_t^{\infty} \frac{d\tau}{a(\tau)} \,. \tag{10.12}$$

Galaxies currently further from us than $d_{EH}(t_0)$ are no longer in causal contact with us and never again shall be: they are, in effect, no longer visible. Of course, we may still be able to 'see' them but only as photons emitted in the past: it is not possible in principle to observe anything beyond our event horizon as it currently exists, no matter how long we wait.

To see which universes have finite event horizons, change the variable of integration from t to a in the above equation to derive an alternative expression for the current time:

$$d_{EH}(t_0) = c \int_1^{\infty} \frac{1}{a} \frac{da}{\dot{a}(a)} \,. \tag{10.13}$$

For the integral to converge at its upper limit and the event horizon to be finite, $\dot{a}$ must be positive definite ($\dot{a} > 0$ always) and a non-decreasing function of a; i.e., $\ddot{a} \geq 0$. Only non-decelerating models can have finite event horizons, all others have infinite event horizons and represent universes that forever remain causally connected throughout their particle horizons.

The flat, matter-only model is non-accelerating and has an infinite event horizon; but the de Sitter model, with $a(t) = \exp[H_0(t - t_0)]$, has a finite and constant event horizon $d_{EH} = c/H_0 \approx 4.15$ Gpc (for the canonical H_0). Models containing both matter and cosmological constant can display rather complicated horizon evolutions; see Figure 15.8 as an instructive example.

10.3.5 Others

Whatever the particle or event horizons, we cannot actually see anything behind the CMB wall corresponding to the time of photon decoupling (technically, the last scattering surface). This **visibility horizon** delimits the extent of the currently visible Universe. In all likely models it corresponds to $z \approx 1100$ and is somewhat less than the current particle horizon: we cannot currently see those things that have influenced our current Universe, but occurred before the era of photon decoupling. In our flat, matter-only example model, for instance, $d_{\mathrm{PH}} = 8.297$ Gpc while the visible horizon distance is only $d_{\mathrm{VH}} = 8.046$ Gpc (both for $H_0 =$ 72 km/sec/Mpc $= 0.074$ Gpc^{-1}).

Under some circumstances we are concerned with pressure wave propagation where the effective causal barrier is set by the speed of sound. From Equation (9.53) the proper distance to a co-moving object evolves with expansion as $\dot{d}_{\mathrm{p}} = H d_{\mathrm{p}}$, so the **sound horizon** – the furthest sound can travel from its time origin – is given by Equation (10.6) with v set equal to the speed of sound c_{s}:

$$d_{\mathrm{SH}}(t) = a(t) \int_0^t \frac{c_{\mathrm{s}}(\tau)}{a(\tau)} d\tau \,. \tag{10.14}$$

The limits of integration can be varied as needed. See Section 17.3.1 for an application of the sound horizon concept, which can readily be extended to other spatial velocities.

10.4 Expansion realities

10.4.1 Galaxy visibility

The visibility of the Universe's galaxy population is a trade-off between growths of horizons and of proper distances. For the flat, matter-only model proper distances to galaxies grow as $a(t) \propto t^{2/3}$ while the particle horizon grows as $d_{\mathrm{PH}}(t) \propto t$; so the number of galaxies within the horizon grows with time. But in a flat, Λ-only model both proper and particle horizons grow as $a(t) \propto \exp(H_0 t)$ so the contents of the particle horizons are unchanging, while the event horizon remains static at $d_{\mathrm{EH}} = c/H_0$: the number of galaxies within the event horizon – the number remaining in causal contact – thus decreases exponentially with time in such accelerating models. It is in this sense that accelerating universes thin out over time.

The matter is illustrated in Figure 10.4 in which galaxy distances and horizons are plotted for two times differing by a factor of 2 in their expansion parameter a. The number of galaxies (dots) within the particle horizon increases with time in a matter-only (or matter+radiation) model, but remains constant in Λ-only models. The number of galaxies within the event horizon (if there is one) decreases with time.

Figure 10.4 illustrates another aspect of the visible Universe. In practice we cannot achieve such a 'God's-eye' view of the current Universe, and galaxies near

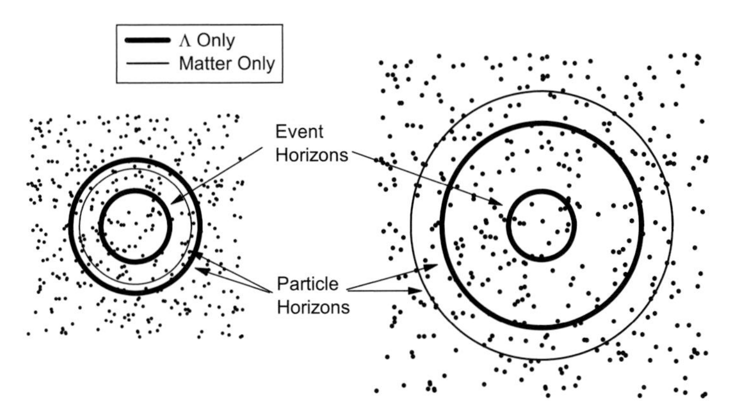

Figure 10.4 Simulated galaxy distribution in evolving flat, Λ-only models (heavy lines) and flat, matter-only models (light lines): the two panels differ in time by a factor of 2 (left panel = earlier) in their expansion factors. Inner circle: event horizon. Outer circles: particle horizons.

our particle or visibility horizon are seen as they were in the past, or even before they had fully formed. The limitation of our view to within horizons is a temporal one, and perfectly good galaxies currently exist beyond the horizon even if we cannot see them. Thus, the galaxy distributions shown in Figure 10.4 can extend to infinity, as far as we can tell; but we cannot see them beyond our horizon.

10.4.2 What's expanding?

Universal expansion is commonly viewed as expansion of space itself, which seemingly entails expansion of everything in the Universe. This would presumably lead to expansion of things like the Solar System, the Earth itself, meter sticks – and even your body, dear reader. This 'expanding space' interpretation is eerily reminiscent of the Luminiferous Aether invoked a century or more ago to explain the propagation of electromagnetic waves in the era prior to quantum field theory: some mysterious property of space itself provides the requisite mechanism. But pretty clearly this is not happening with universal expansion: even if expanding meter sticks would leave measured dimensions of expanding objects unchanged, the dynamical consequences of such all-encompassing expansion would be substantial and easily detected – but they aren't, at least not locally (see Problem 3 of this chapter).

What is actually happening in our physical expansion picture is that everything in the Universe is falling freely in the universal gravitational field, unless acted upon by some other force: in this picture, *space* is not expanding but our co-moving metric is. Objects are not pulled along by 'expanding space', but fall away from each other unless some other force holds them together: local gravitational forces in the case of Solar System orbits, mechanical forces in the case of meter sticks, etc.

We can estimate the extent to which local forces appear to counteract universal expansion by comparing expansion and contraction rates. A force-free structure of length L will have its size increased by universal expansion at rate $\dot{L} = \dot{a}(t_0) L = H_0 L$. The appropriate expansion time scale is thus $L/\dot{L} = 1/H_0 = t_H$, the Hubble time ($\sim$ 13.5 Gyr or 4×10^{17} seconds for our canonical H_0). On the other hand, the structure's gravitational contraction time scale is approximately[2]

$$t_{gc} \approx (G\rho_m)^{-1/2} . \tag{10.15}$$

For contraction to just balance expansion, $t_H \sim t_{gc} \Rightarrow \rho_{m,0} \approx 10^{-27}$ kg/m^3, on the order of the mean matter density of the current Universe. It follows that a self-gravitating structure of mass density greater (approximately) than that of the overall Universe will currently be able to contract under gravity even as the Universe expands. Since your body is about 25 orders of magnitude more dense than is the entire Universe, you are in no danger of being expanded. One can easily show that the same is true for nearly all self-gravitating structures in the Universe other than the very largest: universal expansion, to first order, does not internally expand bound structures. Problems at the end of the chapter invite you to show that the densities of gravitationally bound structures by and large vary with structure mass in an inverse manner: the most massive structures (e.g., galaxy superclusters) appear to be the weakest bound as measured by their mean densities.

But while universal expansion may not always prevent gravitational compression and binding, it can certainly slow or weaken it, so that dynamical analyses of structure growth in an expanding Universe must take cosmological expansion into account. This is a difficult matter that we discuss briefly in Chapter 18. As a simple example, our Local Group of galaxies contains a few dozen galaxies spread over a roughly spherical volume of $\sim$ 1 Mpc diameter. It appears to be contracting: the largest member, M31 (the Andromeda spiral), is moving toward us at better than 100 km/sec, as are several other medium-to-large members (all the negative radial velocities shown in Figure 10.4 are Local Group members). A plausible scenario is that all these galaxies were originally moving apart when the Universe was young, but have turned around as their mutual gravitation has stopped the expansion.[3] The same may be true for all or nearly all galaxy clusters and superclusters.

In a flat, matter-dominated Universe (which ours has nearly been for most of its galaxy-forming history), $a(t) \propto t^{2/3}$ and the Hubble time goes as $t_H \propto t$, while the mean density is $\rho \propto a^{-3} \propto t^{-2}$ so $t_{gc} \propto t$ also; the ratio t_H/t_{gc} of expansion to

[2] A standard result in star formation theory; see, e.g., Section 12.2 of Carroll and Ostlie (2007).
[3] See, e.g., Section 4.5 of Sparke and Gallagher (2007) for a non-relativistic discussion of the evolution of dynamics of the Local Group.

contraction times thus does not change as the Universe expands. The fact that we currently see structures on nearly all scales then suggests that such structures are stable and will not be disrupted as universal expansion proceeds. But, once again, dark energy is a potential spoiler: its consequent exponential universal expansion may disrupt existing structures (particularly the largest such) in the future.

10.4.3 Expansion velocities

Consider a galaxy whose current proper distance is $d_0 = 4.8$ Gpc. From Equation (9.52), expansion is increasing this distance at the rate $\dot{d}(t) = \dot{a}(t)\, d_0$, or $\dot{d}(t_0) = H_0 d_0 = 0.355$ Gpc/Gyr, about $1.16c$. The galaxy currently appears to be receding from us at greater than the speed of light, all the while its emitted photons are making their way to us.

This is a common occurrence in all realistic cosmological models based on the RW metric: since $\dot{a}$ becomes arbitrarily large at early times in Big Bang models; and d_0 is arbitrarily large for distant galaxies in an unbounded Universe; apparently superluminal velocities are an unavoidable fact of life for cosmologists. There is no general consensus as to how to interpret these velocities, but most attempts to do so involve one or another of the following approaches.

First, the numerical value computed for such velocities, $\Delta d/\Delta t$, is partly a consequence of the choice of coordinate systems, and the system used in the RW metric is neither unique nor entirely straightforward. Consider, for instance, the empty model $\dot{a}^2 = -K_0 c^2$ of Section 8.6, for which a solution is $a(t) = \sqrt{-K_0}ct$ where $K_0 \leq 0$ is an arbitrarily chosen number defining the metric system employed. From Equation (9.47) the proper distance to radial coordinate r_g is then

$$
\begin{aligned}
d_p(r_g, t) &= c\, t \operatorname{arcsinh}\left(\sqrt{-K_0} r_g\right) , \\
\Rightarrow \dot{d}_p &= c \operatorname{arcsinh}\left(\sqrt{-K_0} r_g\right) .
\end{aligned}
$$

Sufficiently remote places, for which $r_g \gtrsim 1.175/\sqrt{-K_0}$, have $\dot{d}_p > c$ at all times. But K_0 is entirely arbitrary in this model; in particular, the expansion equation for the empty model can be satisfied with $K_0 = 0$ so that $\dot{d}_p = 0$ and there is no expansion at all, let alone a superluminal one. Expansion velocities in this model are entirely metric in nature, and have no physical meaning.

Admittedly, this is an extreme example; but it serves to demonstrate that velocities computed on cosmological scales can reflect the choice of metric as well as the underlying reality. This arbitrariness of choice of metric and consequent expansion rates extends to more realistic models containing gravitating components, if to a lesser extent. In particular, different choices of the cosmological time coordinate can eliminate many of the superluminal velocities encountered in

RW models. The physical meaning of velocities computed with metric quantities arising from co-moving coordinates thus appears to be rather dubious.

Another approach to understanding superluminal velocities in cosmology is to abandon the sanctity of constancy of the speed of light, and of its limiting nature. This is not as heretical as it may seem at first, for the familiar properties of c derive from electromagnetism in inertial reference frames and may not hold in cosmological contexts. Indeed, in the opinion of some (e.g., Graves 1971) the types of coordinate system employed in GR compel a non-constant speed of light, and a role for c that is principally one of a conversion factor between spatial and temporal quantities. Section 21.1.2 of Ellis *et al.* (2012) discusses some other possibilities for a non-constant speed of light in cosmological contexts.

A third approach is to invoke expansion of space itself, which is supposedly unrestrained by such things as the speed of light and simply carries galaxies along with it. For reasons discussed in Section 8.6 this seems to be something of a cop-out and serves mostly as a metaphor – admittedly a harmless one in many instances. If nothing else it has the virtue of simplicity, but perhaps one should invoke it only within quotation marks.

The best way to deal with the matter of superluminal velocities in cosmology is probably to view proper distances, and velocities computed therefrom, as *metric* quantities that reflect the chosen coordinate system as well as physical realities. Since velocities computed in the local limit (Δt, $\Delta r \to 0$) obey $\left|\Delta d_{\mathrm{p}}/\Delta t\right| \leq c$ by virtue of the definition of proper distance, incremental velocities are physically meaningful. But any velocity computed over a time interval that is not very short in comparison with the Hubble Time should be interpreted with care.

Problems

1. In our current model for the Universe (the CCM of Chapter 15), between the times of radiation-matter energy densities equality $\left(a \approx 3 \times 10^{-4}\right)$ and that of photon decoupling $\left(a \approx 10^{-3}\right)$ matter dominated the expansion and its density was $\Omega_{\mathrm{m},0} \approx 0.3$. Show that the Universe was dynamically flat during that time also, as it was when radiation dominated; so the Universe was effectively flat up to the time of photon decoupling.
2. Derive an expression for the event horizon for expansion models of the form $\alpha(t) \propto t^x$ for $0 \leq x < 1$. Make a plausibility argument that horizons should vary with expansion rates in this manner. Comment on the pathological case $x = 1$, and the cause of such behavior.
3. Assume that the Sun–Earth system is expanding with the Universe's co-moving coordinate system, and compute the relative change in Sun–Earth distance over the course of, say, the last billion years during which life was evolving on the Earth. Comment on the consequences.

4. Estimate the mean mass densities of gravitationally bound structures in the Universe of different masses: the Earth–Moon system; the Solar System; the Milky Way galaxy; the Local Group of galaxies; the Virgo Cluster of galaxies; and a representative galaxy supercluster. Comment on the systematics of your results in terms of gravitational binding against universal expansion.

Part IV

Expansion models

Solutions to the Friedmann Equations are models of the Universe's expansion and constitute a four-parameter family of expansion functions $a(t)$; it is the cosmologists' job to determine which of them is correct, or at least plausible. We approach this problem from two directions: estimations of the densities of sources of gravitation from observations, for use in solving the Friedmann Equations; and comparisons of predictions of trial solutions with observed structure and kinematics of the Universe's contents.

We begin by solving the Friedmann Equations for single-component models, and demonstrating the comparison of their predictions with observations of the real Universe. The best-fitting model at present is the Concordance Cosmological Model (CCM), containing radiation, matter, and dark energy; and discussed in detail in Chapter 15.

11

Radiation

That radiation is a source of gravitation is a consequence of the special relativistic equivalence of matter and energy. Radiation is commonly overlooked in this regard because its energy density in the current Universe is typically very small. The densest form of radiation in our vicinity, for instance, is solar irradiance at $\mathcal{F} \approx 1350$ W/m^2 (at the Earth's distance from the Sun), corresponding to an energy density of $\varepsilon \approx \mathcal{F}/c \approx 4.5 \times 10^{-6}$ J/m^3 or an equivalent mass density of $\rho = \varepsilon/c^2 \approx 5 \times 10^{-23}$ kg/m^3. This is thinner than the 'vacuum' between planets, or between stars in our part of the Galaxy. Under most non-cosmological circumstances there is thus no need to include radiation in computing gravitational forces in the current Universe. But in the very early Universe radiation was the dominant form of gravitation and forms the basis of expansion models prior to $t \sim 10^4$ years, or about one millionth of its current age.

For cosmological purposes it is useful to describe 'radiation' as highly relativistic particles, for which the EOS parameter is $w = 1/3$ and $\varepsilon = \varepsilon_0 a^{-4}$. In the very early Universe this was true of many elementary particles; in the current Universe it applies only to photons and to light neutrinos. Current photon sources are many and varied, producing measurable radiation energy densities at wavelengths from gamma to radio waves.[1] Of these types of radiation two are the largest contributors to the overall radiation density: optical/IR radiation, mostly from stars and stellar systems, with $\varepsilon_{\text{opt}} \approx 10^{-15}$ J/m^3; and microwaves with $\varepsilon_{\text{micro}} \approx 6 \times 10^{-14}$ W/m^3. Almost all this microwave energy is in the form of a nearly uniform and isotropic radiation field called the Cosmological Microwave Background Radiation (CMB), which is thus the dominant form of radiation in the current Universe. The current CMB energy density is nearly two orders of magnitude greater than the combined densities of all other forms of radiation, but is still more than four orders of magnitude less than the energy density of (non-relativistic) matter.

[1] See table 9.1 of Longair (2008) for current radiation energy density accountings.

11.1 Cosmological Microwave Background Radiation

The **Cosmological Microwave Background Radiation (CMB)** is a remnant of the hot, dense early Universe. The origin of the CMB appears to be a thermal equilibrium between radiation and matter during the period when the Universe's contents were ionized and photons scattered readily off free electrons, producing a nearly uniform and isotropic radiation field of the same temperature as baryonic matter. Observed today, the CMB radiation is a ubiquitous microwave background whose spectrum, with an intensity maximum of a few times 10^{-18} W/m^2/Hz/sr at a wavelength of $\sim$ 2 mm, corresponding very closely in both amplitude and spectral form to a black-body of temperature 2.725 ± 0.001 K: there seems little room for doubt concerning its black-body nature (see Figure 11.1). This background radiation parted company from matter at the time of photon decoupling – commonly (if confusingly) referred to as 're-ionization' – when electrons were captured onto positive ions, a consequence of the expanding Universe having cooled to $\sim$ 3000 K, and photon scattering largely ceased. Since that event CMB photons have been mostly free to expand throughout the Universe without being scattered, and are observed today as the CMB radiation.

It is easy to show that uniform stretching of all black-body photon wavelengths, as would be produced by redshifts arising from the Universe's expansion, preserves the black-body intensity spectrum of radiation at a temperature reduced by the same expansion factor: $\lambda \rightarrow \lambda \times a \Rightarrow T \rightarrow T/a$. The energy density of black-body radiation of temperature T is $\varepsilon_{\mathrm{r}} = a_{\mathrm{rad}} T^4$, where $a_{\mathrm{rad}} = 7.566 \times 10^{-16}$

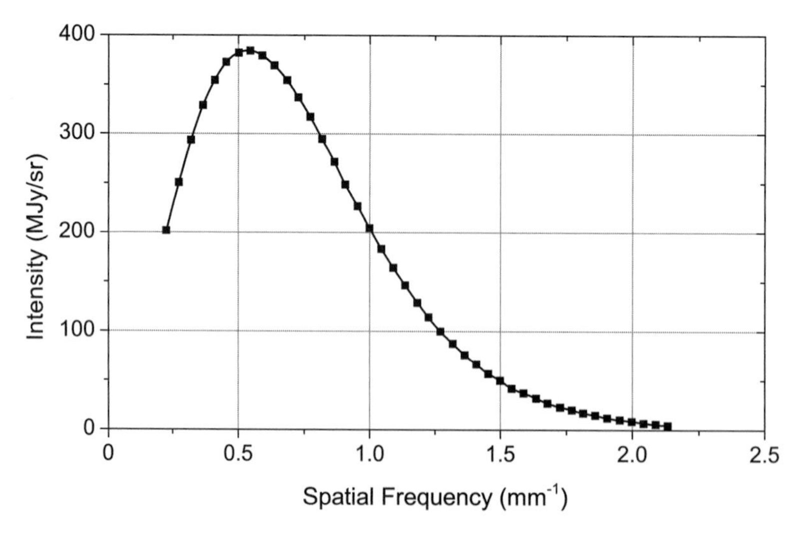

Figure 11.1 Cosmological Microwave Background Radiation spectrum. Points: values observed by the COBE satellite (error estimates smaller than the plotted symbols). Line: theoretical spectrum from a black-body of temperature 2.725 K.

J/m^3/K^4 is the radiation constant. Then from Equation (8.29) for radiation energy density in an expanding Universe, $\varepsilon_r \propto a^{-4}$ so the CMB temperature varies with expansion as

$$T_{\text{CMB}}(a) = \frac{T_0}{a}, \tag{11.1}$$

where T_0 is the current temperature. Since $1 + z = a^{-1}$, the CMB temperature corresponding to redshift z is

$$T_{\text{CMB}}(z) = T_0 \times (1 + z) \ . \tag{11.2}$$

Photon decoupling was at $T \approx 3000$ K, so the current 2.725 K corresponds to a CMB redshift of $z_{\text{CMB}} \approx 1090$. The CMB photons currently carry mean energies of $\langle \mathcal{E}_0 \rangle \approx 2.70\, k_B T_0 \approx 0.6$ milli-eV per photon and are capable of energizing very low-lying energy levels in some interstellar molecules, with subsequent radiative decays that produce observable line radiation. This radiation has been observed both in our Galaxy and (in CO molecules) in intergalactic clouds of redshift $z \approx 2.4$, where its intensity corresponds to $T \approx 9$ K in good agreement with Equation (11.2); again confirming the black-body nature of the CMB.

The current energy and photon number densities of the CMB are given by the appropriate black-body relations:

$$\varepsilon_{\gamma,0} = a_{\text{rad}} T_0^4 = 4.17 \times 10^{-14} \text{ J/m}^3 ,$$
$$n_{\gamma,0} = \beta T_0^3 = 4.11 \times 10^8 \text{ photons/m}^3 ,$$

where $\beta = 2.03 \times 10^7$ m^{-3}K^{-3}. By comparison, the current mean energy density of visible/IR radiation in the Universe is nearly two orders of magnitude less than that for CMB radiation, and the photon number density is $\sim 10^4$ photons/m^3, four orders of magnitude less than the CMB.[2] The mean CMB photon flux is currently $\mathcal{F} = n_{\gamma,0} \times c/\sqrt{3} \approx 7 \times 10^{16}$ photons per second per square meter; the high-energy tail of the CMB photon energy distribution shows up as noise ('speckles') on analog televisions tuned off-channel.

The total mass–energy density of the Universe is the near equivalent of the rest mass of six baryons per cubic meter, a number density that is eight orders of magnitude less than that for CMB photons. The baryon number density is a small fraction of that for all forms of matter, amounting to less than one baryon per cubic meter; so that photons outnumber baryons by about nine orders of magnitude. This is normally expressed as the **baryon/photon number ratio**:

$$\eta \equiv \frac{\text{number density of baryons}}{\text{number density of photons}} \sim 10^{-9}. \tag{11.3}$$

[2] Longair (2008), table 9.1.

As we shall argue below, this number is nearly constant during the Universe's expansion and is a fundamental characteristic of our Universe (a more precise estimate of its value is given in Section 16.3). It is fair to view the Universe as a sea of photons (and neutrinos; see below), speckled here and there by the occasional proton/neutron/electron. This imbalance between photons and baryons is of some importance in understanding the evolution of the Universe and is discussed further in Chapter 16.

11.2 Neutrinos

The current Universe also contains primordial neutrinos left over from early times when energies were high enough to support neutrino coupling to baryons and leptons via reactions mediated by the weak nuclear force, such as

$$\mathrm{n} + \nu_{\mathrm{e}} \rightleftarrows \mathrm{p} + e^{-} \ ,$$
$$\mathrm{p} + \bar{\nu}_{\mathrm{e}} \rightleftarrows \mathrm{n} + e^{+} \ .$$

These reactions effectively stopped as the reaction rates fell below the Hubble rate H when mean particle energies fell below a few MeV; which occurred at $t \sim 1$ second. From that point on these neutrinos were decoupled from both radiation and matter in the Universe, and exist today as a primordial neutrino background.

At about $t \approx 30$ sec, when mean particle energies fell below the electron rest energy of $\approx \frac{1}{2}$ MeV, electron–positron pair production stopped and the remaining positrons annihilated with electrons to produce many of the Universe's photons; in the process the Universe was brought to a higher temperature than that of the primordial (and decoupled) neutrinos. Detailed calculations based on entropy conservation predict $T_{\mathrm{neutrinos}} = (4/11)^{1/3}\, T_{\mathrm{CMB}} \approx 0.71\ T_{\mathrm{CMB}}$ and an energy density of all three neutrino families combined of[3]

$$\varepsilon_{\nu} = 3\frac{7}{8}\left(\frac{4}{11}\right)^{4/3} \varepsilon_{\mathrm{CMB}} = 0.68\ \varepsilon_{\mathrm{CMB}} \ .$$

If neutrinos are massless they will remain relativistic throughout the Universe's history, and this energy imbalance between primordial neutrinos and photons will remain unchanged. The result would be a current radiation energy budget 68% higher than that of CMB radiation alone, or

$$\varepsilon_{\mathrm{r},0} \approx 1.68 \times a_{\mathrm{rad}} T^{4}_{\mathrm{CMB},0} \approx 7.01 \times 10^{-14}\ \mathrm{J/m^3} \ .$$

This is what is normally taken as the current radiation energy density.

[3] See, e.g., Section 9.6.5 of Ellis *et al.* (2012), or Section 5.3 of Narlikar (1993).

Evidence for a non-zero neutrino mass has been gathering for the last two decades, so that neutrino rest masses of $\lesssim 0.1$ eV seem possible and thus the primordial neutrinos are probably non-relativistic at present. With a mass of ~ 0.1 eV, a primordial neutrino would have become non-relativistic well after photon decoupling, when matter dominated the expansion and matter densities were much greater than that of any plausible neutrino density; so it does no harm to include primordial neutrinos – massive or not – in the Universe's radiation budget as if they were relativistic, and ignore them in the matter budget where the issue of their relativistic nature is moot. This is what is done in the CCM model to be introduced in Chapter 15.

11.3 Entropy

From the first law of thermodynamics with a constant number of particles, $dE = T\,dS - P\,dV$, so

$$\frac{dS}{dV} = \frac{1}{T}\left(\frac{dE}{dV} + P\right) .$$

For black-body radiation both E and S are extensive thermodynamic quantities; i.e., they are proportional to V so that $dS/dV = S/V$, and similarly for E. Then the above equation becomes

$$s = \frac{\varepsilon + P}{T} \quad \text{(black-body radiation),} \tag{11.4}$$

where $s \equiv S/V$ is the entropy density.[4] Now, for radiation $\varepsilon + P = (4/3)\,\varepsilon \propto a^{-4}$, while from Equation (11.1) $T \propto a^{-1}$ and thus the above relation implies $s \propto a^{-3}$, or $S \propto sa^3 = \text{constant}$.

The CMB entropy remains constant as the Universe expands.

It is instructive to compute the current value of this density for comparison with that of non-relativistic matter: from $\varepsilon_{\text{rad}} = a_{\text{rad}}T^4$ and $P = \varepsilon/3$, the above equation for entropy density becomes

$$s_{\text{CMB}}\,(t_0) = \frac{4}{3} a_{\text{rad}} T_0^3 \approx 2 \times 10^{-14}\ \text{J/K/m}^3 ,$$

for a current temperature of $T_0 = 2.725$ K.

[4] See chapter 4 of Bernstein (1995) for a detailed and rigorous proof of this result.

By comparison, the entropy density of weakly interacting gas particles is on the order of $s_{\text{gas}} \sim nk_{\text{B}}$, where k_{B} is Boltzmann's constant and n is the gas particle number density. For the current estimated mass density of the Universe, $\rho_{\text{m}} \approx 0.27\rho_{\text{crit},0} = 2.7 \times 10^{-27}$ kg/m^3 (including dark matter), or the rough equivalent of $n \sim 1$ baryon per cubic meter; so the current entropy density of non-relativistic gas is on the order of

$$s_{\text{gas}}(t_0) \sim k_{\text{B}} \approx 10^{-23} \text{ J/K/m}^3 ,$$

or about nine orders of magnitude less than that of the CMB. Condensed objects add some to this entropy accounting, but not much: most of the non-relativistic matter in the Universe is probably in the form of gas, and stars have entropies that do not differ much from gas of the same mass;[5] so the total matter entropy of the Universe is nearly nine orders of magnitude less than that of the CMB. This is the basis for the impression that the entropy of the Universe remains constant as it expands: the entropy of the matter content of the Universe may (and certainly does) increase with time, but such increases are totally swamped by the much larger and invariant primordial radiation entropy. To first order, the Universe's entropy remains constant.

The entropy of both radiation and weakly interacting matter is given approximately by $s \sim k_{\text{B}}n$ so that the ratio of number densities of CMB photons to baryons is approximately 10^9, and is nearly constant. Then the baryon/photon number ratio (Equation (11.3)) is, to first order, a universal constant of value $\eta \sim 10^{-9}$ (we shall refine this estimate in Chapter 16).

11.4 Radiation-only expansion models

If only radiation is present the Expansion Equation (8.32) becomes

$$\dot{a}^2 = H_0^2 \left[\frac{\Omega_0}{a^2} - (\Omega_0 - 1) \right] , \tag{11.5}$$

to which the solution for $a(0) = 0$ is

$$a(t) = \left[2H_0\sqrt{\Omega_0}t + H_0^2 (1 - \Omega_0) t^2 \right]^{1/2} \quad \text{(radiation only).} \tag{11.6}$$

This represents eternally expanding universes for open or flat geometries ($\Omega_0 \leq 0$); for closed geometries the model stops expanding at some point and recontracts thereafter.

[5] Excluding the entropy residing in black holes, which may (depending upon the size and contents of the Universe's black hole population) be greater than that of the CMB (see Problem 1).

From Section 10.1 we expect the early Universe to be radiation-dominated and effectively flat up to the time of equality of radiation and matter energy densities which – depending upon the specific model adopted – probably occurred at an age between $\sim 10^4$ and $\sim 10^5$ years. Thus, the only useful 'radiation-only' models are for the early Universe when $H_0 t \ll 1$, and so take the form

$$a(t) = \left(2H_0\sqrt{\Omega_{\mathrm{r},0}}\right)^{1/2} t^{1/2} \quad (\text{early; } t \ll 1/H_0)\,, \tag{11.7}$$

for initial condition $a(0) = 0$. Given the well-determined value $\varepsilon_{\mathrm{r},0} = 7.01 \times 10^{-14}$ J/m^3 implied by the CMB temperature, and the canonical $H_0 = 72$ km/sec/Mpc $= 0.074$ Gpc^{-1}, $\Omega_{r,0} \approx 8.0 \times 10^{-5}$ and thus

$$a(t) \approx 2.05 \times 10^{-10}\, t^{1/2}\,, \tag{11.8}$$

for t in seconds during the radiation era, which extends from the end of inflation ($\sim 10^{-34}$ sec?) up to nearly the onset of matter dominance. In all currently plausible cosmological models, this is the appropriate expansion function to be used in the early Universe where such interesting things as nucleosynthesis are taking place.[6] The Universe's expansion rate in these early times is

$$H(t) = \frac{\dot{a}(t)}{a(t)} \approx \frac{1}{2t}\,, \tag{11.9}$$

and the Hubble Time is

$$t_{\mathrm{H}}(t) \equiv \frac{1}{H} \approx 2t\,. \tag{11.10}$$

This is, roughly, the time needed for the radiation-dominated Universe to double in size. The particle horizon for a flat, radiation-only model grows as (Equation (10.9))

$$d_{\mathrm{PH}}(t) = 2ct \quad (\text{flat, radiation only}). \tag{11.11}$$

Problems

1. The entropy of a non-rotating black hole of mass M is

$$S_{\mathrm{blackhole}} = 8\pi^2 \frac{Gk_{\mathrm{B}}}{hc} M^2\,,$$

6 Modified somewhat prior to $t \approx 30$ seconds by non-adiabatic events such as pair production and particle freezeouts; see Chapter 16.

where k_B is Boltzmann's constant and h is Planck's constant. Calculate the entropy of the putative 4.1-million solar mass black hole at the center of the Milky Way galaxy. Assuming that most of the Universe's baryons are in similar galaxies, each of them harboring a similarly sized black hole; make a rough estimate of the mean black hole entropy density of the Universe and compare it to the CMB entropy density calculated in Section 11.3.3.

2. Find expressions for the current ages of radiation-only models in terms of H_0 and Ω_0. Use these to show that, for a given value of H_0, more open models are older.
3. Graph the expansion function for closed, radiation-only models and demonstrate that such models eventually stop expanding and recontract. Find an expression for the turn-around time of a closed, radiation-only model; i.e., the time at which expansion stops and contraction sets in.

12

Matter

From a cosmological perspective, **matter** is the energy component with negligible EOS parameter so that its energy density varies with expansion as $\varepsilon_{\mathrm{m}} \propto a^{-3}$. Unlike the case with radiation, matter in the Universe comes in many forms and is only partially visible or otherwise detectable, so its gravitating density is difficult to evaluate.

The current standard model of particle physics includes two main types of fermion, *hadrons* and *leptons*. Hadrons are quark composites and are the only particles subject to the strong nuclear force. They are further divided into two types. **Baryons** are three-quark composites: the only stable baryons are protons and neutrons,[1] so that atomic nuclei are made up entirely of baryons, and protons/neutrons are thus often called **nucleons**. **Mesons** are two-quark composites and are all unstable on short time scales, so that the only stable hadrons are the nucleons. **Leptons** are subject to electroweak forces but not to the strong nuclear force; stable leptons are the electron and its anti-particle the positron, and neutrinos.[2] For convenience – and since the Universe is overall electrically neutral – electrons are often included with nucleons as 'baryonic matter'.

Baryonic matter is what is usually meant by 'normal matter'– it is the stuff we, and all we can see around us, are made of. But our ability to detect fundamental particles is limited to masses less than those achievable by current particle accelerators, which (with the advent of the Large Hadron Collider) are limited to masses on the order of a few times 100 GeV or less (nucleons have masses of a bit less than 1 GeV). Since some exotic particle theories predict particles of even

[1] Neutrons are stable only when bound into nuclei with protons; some current particle theories predict that the proton is also unstable, but with a half-life so long that protons are effectively stable for cosmological purposes.

[2] Neutrinos are subject to oscillations between types (electron, muon, tauon), but probably do not decay to non-neutrinos.

greater mass, there is probably a lot left to discover in the realms of possible forms of matter.

So it really should be of no great surprise to find cosmological evidence for currently unknown forms of matter. The gravitational evidence for 'dark matter' has been growing for at least 80 years, and has reached the point of near certainty. This exotic stuff appears to have no appreciable cross-section to electromagnetism so that it neither absorbs nor emits detectable radiation: hence, 'dark'. It can only be detected and measured in terms of the influence of its gravitation on ordinary matter and light.

An accounting of the Universe's matter content – both normal and dark – is given in Sections 12.2 through 12.5. In anticipatory summary: the total amount of gravitating matter is much less than what is needed to close the Universe, and the large majority of matter is dark. But the Universe is so large and sparse that any such accounting must be viewed as preliminary in nature. Before proceeding with our best effort in this regard, the following section sets out the influence of all sorts of matter on the Universe's expansion.

12.1 Matter-only expansion models

If only matter is present the Expansion Equation (8.32) is

$$\frac{da}{dt} = H_0 \left[\frac{\Omega_{\mathrm{m},0}}{a} + 1 - \Omega_{\mathrm{m},0} \right]^{1/2} . \tag{12.1}$$

For flat geometries this reduces to $\dot{a} = H_0 a^{-1/2}$, which integrates trivially to

$$a(t) = (t/t_0)^{2/3} , \qquad t_0 = \frac{2}{3H_0} \qquad \text{(flat, matter-only).}$$

This is the canonical Einstein–de Sitter example used in previous chapters. Proper and luminosity distances are given by

$$d_0(z) = \frac{2c}{H_0} \left[1 - (z+1)^{-1/2} \right] \qquad \text{(flat, matter only),} \tag{12.2}$$

$$d_{\mathrm{L}}(z) = \frac{2c}{H_0} \left[z + 1 - \sqrt{(z+1)} \right] \qquad \text{(flat, matter only);} \tag{12.3}$$

and the Hubble Relation is displayed in Figure 14.4.

For non-flat models the solutions to Equation (12.1) are parametric ones,[3] a cycloid for $\Omega_{m,0} > 1$ (closed) and hyperbolic cycloid for $\Omega_{m,0} < 1$ (open).

[3] See Section 4.3 of Narlikar (1993) for derivation of these equations and guidance for their usage.

$$\text{Closed:}\quad t = \frac{\Omega_0}{2H_0\,(\Omega_0 - 1)^{3/2}}\,(\Theta - \sin\Theta)\ ,$$
$$a = \frac{\Omega_0}{2\,(\Omega_0 - 1)}\,(1 - \cos\Theta)\ ,\qquad 0 \le \Theta \le 2\pi\ . \tag{12.4}$$

$$\text{Open:}\quad t = \frac{\Omega_0}{2H_0\,(1 - \Omega_0)^{3/2}}\,(\sinh\Psi - \Psi)\ ,$$
$$a = \frac{\Omega_0}{2\,(1 - \Omega_0)}\,(\cosh\Psi - 1)\ ,\qquad 0 \le \Psi < \infty\ . \tag{12.5}$$

Examples of these expansion functions are shown in Figure 12.1. Note that more open models are older (for a given value of H_0). As with radiation-only models, closed matter-only models eventually stop expanding and re-contract, and thus induce blueshifts in nearby galaxies: from Equation (12.4) the turn-around occurs for $\Theta = \pi$, or at time

$$t_{\text{turn-around}} = \frac{\pi}{2H_0}\frac{\Omega_0}{(\Omega_0 - 1)^{3/2}}\ . \tag{12.6}$$

The current proper distance for all matter-only models is given by

$$d_0\,(z) = \frac{c}{H_0\sqrt{\Omega_0 - 1}}\arcsin\left\{\frac{2\sqrt{\Omega_0 - 1}}{(z+1)\,\Omega_0^2}\left[\Omega_0 z + (2 - \Omega_0)\left(1 - \sqrt{1 + \Omega_0 z}\right)\right]\right\}\ . \tag{12.7}$$

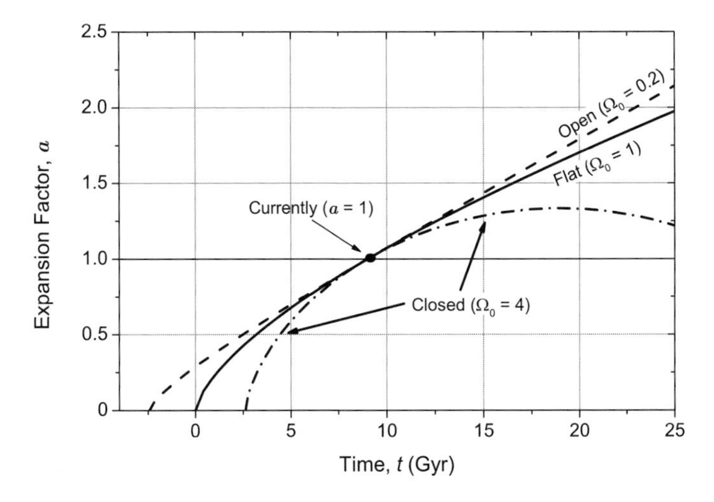

Figure 12.1 Sample expansion functions for matter-only models, all for $H_0 = 72$ km/sec/Mpc. The time axis labels are for the flat model.

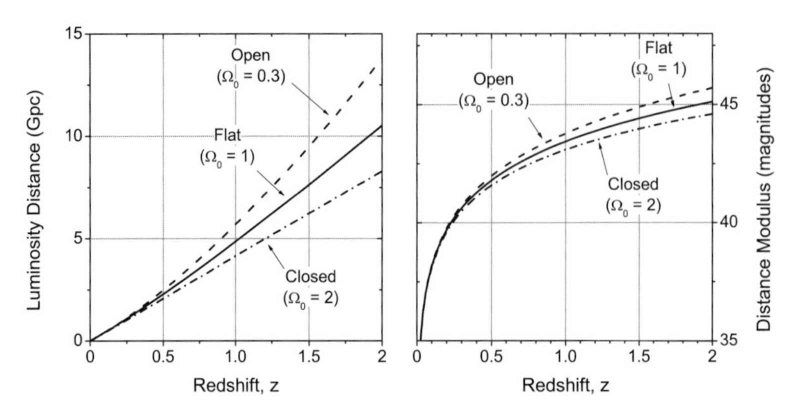

Figure 12.2 Hubble relations for three matter-only models, all for $H_0 = 72$ km/sec/Mpc. Left panel: distance vs. redshift. Right panel: brightness vs. redshift.

Luminosity distances computed from this and Equation (9.61) yield the classic Mattig formula for matter-only models:[4]

$$d_L(z) = \frac{2c}{H_0 \Omega_0^2} \left[\Omega_0 z + (2 - \Omega_0) \left(1 - \sqrt{1 + \Omega_0 z} \right) \right] \quad \text{(matter only)}. \qquad (12.8)$$

Samples of Hubble Relations for matter-only models are shown in Figure 12.2. Note that for given redshift, galaxies are further away and fainter in more open models than in less open ones.

12.2 Baryonic matter

Traditionally, baryonic matter has been measured by its luminosity which is, to some extent, directly observable. Thus, galaxy masses have been estimated by assuming a mass-to-luminosity ratio for the galaxy type, applied so as to infer mass directly from observed luminosity. Mass/luminosity ratios are typically estimated locally, where stars are easily detected and their masses estimated; then applied to more remote galaxies where individual stars and other features are not observable. Thus, the mass/luminosity ratio for our part of the Galaxy is measured to be

$$\left(\frac{M}{L} \right)_{\text{local}} \approx 2 \left(\frac{\mathrm{M}_\odot}{\mathrm{L}_\odot} \right),$$

the value 2 arising from the mix of star types, and amount of interstellar gas, in the solar neighborhood. The luminosity of the similar galaxy M31 is observed to

[4] Peacock (1999), Section 3.4.

be $L_{\mathrm{M31}} \approx 2.7 \times 10^{10}\, L_{\odot}$, so we infer $M_{\mathrm{M31}} \approx 5.4 \times 10^{10}\, \mathrm{M}_{\odot}$ from its luminosity alone. The method can easily be extended to more remote galaxies, provided their luminosities can be measured.

But there is an obvious problem with this approach. As an example, consider the Sun: at a visual absolute magnitude of $M_{\mathrm{V}} \approx 4.8$ its apparent magnitude at the distance of the nearby galaxy M31 ($d \approx 0.7$ Mpc) is $V \approx +29$. This verges on undetectable with modern technology, so that solar-like stars cannot be seen individually at even small cosmological distances. This would not be so worrisome were it not for two characteristics of stellar populations: main sequence luminosities are proportional to a high power of stellar masses, so that most of the Galaxy's luminosity comes from high-mass main sequence stars and giants; and stellar mass distributions in spiral galaxies are strongly biased toward low-mass stars, so most of their mass comes from low-luminosity stars that cannot be seen outside our Galaxy and its satellites. As a result, the detectable luminosities of galaxies typically trace a stellar population that constitutes only a small fraction of galaxy masses. To infer galaxy masses from their luminosities thus requires an extrapolation of what is actually visible, based largely on our understanding of stellar physics and analogies to nearby galaxies. Astronomers have become rather clever at this, but some uncertainty still remains, particularly for galaxy types not well-represented in our vicinity.

For present purposes we define *luminous matter* as that directly inferred from observations of luminous flux, supplemented by extrapolations to such things as stellar remnants and low-luminosity stars and gas, based on galactic models. We observe a mean luminosity density in galaxies of $\mathcal{L}_{\mathrm{galaxies}} \approx 1.5 \times 10^{8}$ solar luminosities per cubic megaparsec (blue light), and we estimate a mean galaxy [luminous mass] to [luminosity] ratio of between 2 and 4 in solar units. The mean density of luminous matter in galaxies is thus $\rho_{\mathrm{galaxies}} \approx 3 - 6 \times 10^{8}\, \mathrm{M}_{\odot}/\mathrm{Mpc}^{3}$. Since the critical mass for closure in these units is $\rho_{\mathrm{c},0} \approx 1.5 \times 10^{11}\, \mathrm{M}_{\odot}/\mathrm{Mpc}^{3}$, the luminous mass density in galaxies is only a tiny fraction of that required to close the Universe:

$$\Omega_{\mathrm{galaxies},0} \approx 0.002 - 0.004 \;\; \text{(stars and gas)} \, .$$

There remain two important non-galactic reservoirs of luminous matter in the Universe. First, groups and clusters of galaxies harbor intergalactic gas that, in many cases, is hot enough to emit detectable X-radiation (cooler gas is mostly non-luminous and cannot be directly detected). In rich clusters such as Coma, the gravitational potential well is so deep that member galaxy velocities are on the order of $\sim$ 1000 km/sec. This helps heat the intracluster gas to a thermally equivalent velocity, typically to several tens of million Kelvins corresponding to particle energies of several keV. Consequently, the gas emits X-rays, probably by thermal bremsstrahlung (free–free emission), so that its mass density may be estimated from the X-ray luminosity and spectral temperature. Typical results for rich clusters is that the mass of the X-ray emitting gas is $\sim$ 5 times that of the luminous mass in member galaxies; the factor is less in poorer clusters, such as Virgo. Integrating this result over the observed number density of galaxy clusters, current estimates of the total amount of hot, intracluster gas in the Universe lie in the range $\Omega_{\text{Intracluster},0} \approx 0.001 - 0.0025$.

A second reservoir is in intergalactic gas at cosmological distances, backlit by (typically) distant quasars: the clouds absorb the illuminating light at the neutral hydrogen resonant Lyman-α line, which is typically seen redshifted to visible wavelengths. Measuring the column density of the clouds from the Lyman-a profiles, and making reasonable estimates for the ionization fraction of the clouds, produces mass density estimates that are comparable to those of galaxies. The method is plagued with two uncertainties: estimates of the ionization fraction of the absorbing gas, and identification of the gas as currently extragalactic. Most of the Lyman-α observations are made of clouds at high redshift (typically $\gtrsim 3$) corresponding to eras during which galaxies may be sweeping up intergalactic gas, so that the gas observed in Lyman-α absorption may be more properly accounted for as intragalactic or intracluster at the current time. As a consequence, mass density estimates corresponding to Lyman-α clouds are uncertain. Sparke and Gallagher (2007) quote $\Omega_{\text{Ly}\alpha} \approx 0.001$, which is probably as good an estimate as any.

The *observable* baryon mass density of the Universe thus probably lies in the range $\Omega \approx 0.004 - 0.008$. But there is good reason to believe that this is much less than the actual baryon mass density: the extent of nucleosynthesis of light elements early in the Universe's expansion is sensitive to the baryon number density, which can thus be estimated by observing current abundances of light elements in environments in which their primordial abundances have been preserved. The matter is discussed in detail in Chapter 16; in summary, our current best estimate for baryon mass density lies in the range[5]

$$\Omega_{\text{baryon},0} \approx 0.038 - 0.045\ . \tag{12.9}$$

Only between $\sim$ 10% and $\sim$ 20% of these baryons are directly observable. It is of some current interest to determine where the 'missing baryons' are sequestered. The most likely candidate is an ionized intergalactic gas: baryons that never made it into galaxies and have been ionized by UV radiation. A sufficiently thin, ionized gas would neither absorb nor emit appreciable radiation and so could go undetected. As could a widely scattered population of intergalactic basketballs, for that matter.

[5] See Fugita & Peebles (2004), *Astrophys. J.* **616**, p. 643; for a detailed accounting of baryon and mass densities in the Universe. Table 7.2 of Sparke and Gallagher (2007) is a partial summary of this accounting.

To carry the Universe's matter accounting further we must extend observations to include non-luminous matter, that observed only indirectly in the form of gravitational effects on luminous matter and radiation. Such accounting is largely of two forms: dynamical mass estimates from observed kinematics of galaxies and clusters, and estimates from gravitational lensing.

12.3 Dynamical mass estimates

The simplest model for the gravitational influence of dark or luminous matter is that of the orbit of a test object about a fixed mass, which may be expressed in terms of centripetal and gravitational accelerations. In a circular orbit these two accelerations are fixed and equal:

$$\mathsf{a} = \frac{V^2}{R} = G\frac{M}{R^2}\,, \tag{12.10}$$

where V is the test object's orbital velocity, R is its orbital radius, and M is the (fixed) central mass. This equality is the basis of Newton's original law of gravitation, and of Kepler's second and third laws of orbital motion. The same equality holds in a radially symmetric system of distributed mass, where now $M = M_{(<R)}$ is the mass interior to the orbital radius. Thus:

$$M_{(<R)} = \frac{R\,V^2\,(R)}{G}\,, \tag{12.11}$$

from which the mass distribution within spherical star systems may be determined from the variation of (circular) orbital velocities with orbital radius.

12.3.1 Disk galaxies

Orbital dynamics are more complicated in non-spherical systems such as the disks of spiral galaxies; but in thin, axisymmetric disks Equation (12.11) remains a good approximation to the relation between rotation velocities and included mass, and is the basis for mass estimates in the disks of spiral galaxies. Outside the central bulge the luminosity densities of spiral galaxy disks typically fall off exponentially with galactocentric distance: $\mathcal{L}\,(R) \propto \exp\,(-R/R_{\rm s})$, where the scale length $R_{\rm s}$ is typically a few kiloparsecs (kpc). If mass follows light so that $\rho\,(R) \propto \mathcal{L}\,(R)$, and the disk is thin, the resulting mass distribution in the disk would go approximately as $M_{(<R)} \propto \int_0^R r \exp\,(-r/R_{\rm s})\;dr$ and, following Equation (12.10), the disk rotation velocity would peak near the scale radius and fall off to larger radii.

But this is not at all what is observed. Figure 12.3 shows *observed* disk rotations for our Milky Way Galaxy, and the similar M31 spiral galaxy, as solid

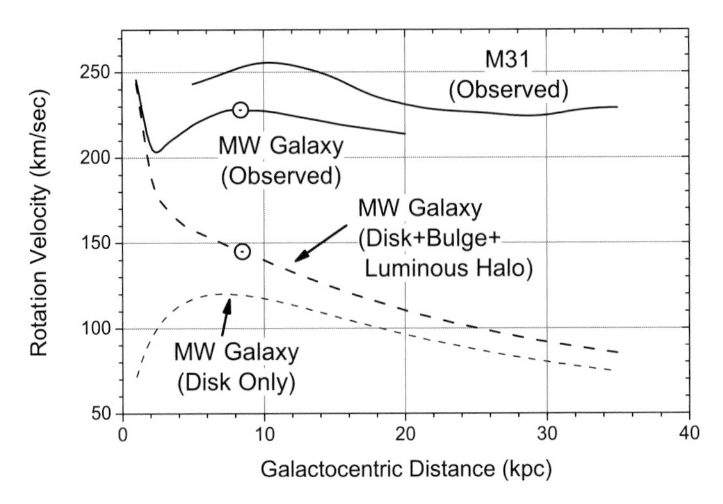

Figure 12.3 Rotation curves in two spiral galaxies. The solid curves trace smoothed observations in the disks of the Milky Way (MW) Galaxy and M31. The dashed curves are the velocities predicted for our Galaxy from observed luminosities of stars and gas, and estimated mass/luminosity ratios. The $\odot$ symbols denote values at the Sun's galactocentric distance.

curves; and the *expected* Milky Way rotation arising from observed luminosities and presumed mass/luminosity ratios as dashed curves. The observed rotation velocities in the Milky Way disk are much larger than those predicted by mass-follows-light modelling; in particular, the Sun's observed galactocentric velocity of $\sim$ 225 km/sec (upper $\odot$ symbol) is greater than the $\sim$ 145 km/sec expected from observed mass densities (lower $\odot$ symbol), implying more mass interior to the Sun's orbit than is observed in luminous matter (stars + gas). Note that reliable tracers of Milky Way rotation well beyond the Sun's galactocentric distance ($\approx$ 8.5 kpc) are hard to come by from our position within the disk, hence the appeal to M31 for rotation velocities beyond the visible limits of spiral galaxy disks.[6] Most large spiral galaxies show the behavior illustrated here by M31: nearly flat rotation curves beyond a few disk scale lengths out to several times the radius of the visible disk, and this is probably true for the Milky Way as well.

From Equation (12.11), a flat rotation curve implies $M_{(<R)} \propto R$ and thus a great deal of dark matter in the outer reaches of these two galaxies. As one might suspect from this figure, galactic rotation velocities imply a much higher mass for the Galaxy than that inferred from its luminous mass (stars + gas) alone. Our best estimates for the Galaxy are a luminous mass (stars and gas) of a few times 10^{10}

[6] See, e.g., Section 2.3 of Sparke & Gallagher (2007) for discussion of the determination of galaxy rotation curves from observations. For the Milky Way Galaxy, the disk scale length is $\sim$4 kpc and the luminous disk radius is $\sim$15 kpc (gas clouds detectable in radio waves extend further).

$M_\odot$, and a total mass interior to the luminous disk limit of $\sim$ 15 kpc – including the dark matter required for the observed rotation velocities – of at least ten times the luminous mass. Dark matter outweighs normal matter in our galaxy (and most other spiral galaxies) by at least an order of magnitude.

Additional evidence for such large amounts of dark matter, both inside and outside the luminous disk limit, comes from the satellite galaxies of the Milky Way. There are at least 12 well-observed such galaxies within $\approx$ 200 kpc of the galactic center, roughly spherically arranged about the Galaxy and all much smaller than the Milky Way or M31. These satellites typically are $\sim 50 - 100$ kpc from the center of the Galaxy and are moving at $100 - 200$ km/sec with respect to the Galaxy as a whole.[7] The escape velocity from the Galaxy's gravitation at galactocentric distance R is $V_{\text{esc}}(R) \approx \left[2GM_{(<R)}/R\right]^{1/2}$ so that $M_{(<R)} \gtrsim \frac{1}{2}V^2R/G$ for these satellites if they are to be bound to the Galaxy, typically implying galactic masses of several times 10^{11} $M_\odot$ at these large distances. For the most part, this is not a convincing argument for individual satellites (for which total velocities usually cannot be determined), but given their numbers and concentrations toward the Galaxy, all implying (or consistent with) similarly large galactic masses, it is likely that many of the presumed satellites are gravitationally bound to the Galaxy, and that a galactic mass of this magnitude is required to bind them.

Whatever the nature of dark matter in spiral galaxies, it appears to mostly reside outside the disk. Stellar motions in spiral galaxy disks are not confined to circular, coplanar orbits, but exhibit oscillatory motion within disk midplane ('epicycles') and perpendicular to it. The frequencies and amplitudes of these motions reveal the mass densities within the disks themselves. The mass distribution of disk stars perpendicular to the disk is approximately exponential: $\rho(z) \approx \rho_0 \exp\left(-|z|/h\right)$, where z is the distance above/below the disk midplane, ρ_0 is the mass density at disk midplane, and h is a scale height characterizing the mass stratification of the disk. Integrating the implied gravitational potentials for thin ($h \ll$ disk diameter), homogeneous disks leads to an acceleration relative to the disk of[8]

$$\begin{aligned} |a_z(z)| &= 4\pi G\rho_0 h\left[1 - \exp\left(-|z|/h\right)\right], \text{ or} \\ \ddot{z} &\approx -4\pi G\rho_0 z, \end{aligned} \tag{12.12}$$

for $|z| \ll h$. This last differential equation is that for harmonic motion, so we expect disk stars near the midplane to bob up and down through the disk with

[7] A listing of satellite galaxies to the Milky Way, together with other members of the Local Group, and relevant physical data, can be found in table 4.1 of Sparke and Gallagher (2007) and table 6.1 of Schneider (2010).

[8] Bertin (2000), Section 14.1. See also Sparke & Gallagher (2007), Section 3.4.1, for a sophisticated analysis employing the Collisionless Boltzmann Equation.

angular frequency $\omega \approx \sqrt{4\pi G\rho_0}$. This density-dependent frequency can be compared to that actually observed in stellar motions in order to estimate disk mass densities. In application to the solar vicinity, this sort of analysis predicts galactic disk[9] mass densities of $\approx 10^{-20}$ kg/m^3 ≈ 0.15 M$_\odot$/pc^3 at the solar galactocentric distance, somewhat greater than what is observed in luminous stars (and their remnants) and interstellar gas; but not nearly enough to account for the dark matter revealed by rotation curves. Dark matter appears to be a minor component of the disk mass of spiral galaxies; most of it resides outside the disk.

Given the extent to which flat galaxy rotation curves extend, it is likely that dark matter in spiral galaxies occupies a roughly spheroidal volume enveloping the 'normal matter' galaxy. In the case of the Milky Way Galaxy, velocities of satellite galaxies suggest that this volume extends to at least the Magellanic Clouds at ~ 50 kpc, and perhaps as far as ~ 200 kpc. This dark matter component of spiral galaxies is typically known as the 'dark halo' or 'corona'. The mass of the Galaxy out to ~ 200 kpc – a significant fraction of the distance to M31, and four times the distance to the Magellanic Clouds – is thus probably $\sim 10^{12}$ M$_\odot$, implying $M/L \gtrsim 50$. From dynamical evidence such as described here, the matter content of spiral galaxies appears to be dominated by dark matter, which probably constitutes $\gtrsim 90\%$ of the total in late-type spirals and exists largely in a spheroidal corona much larger in diameter than the luminous disk. To the limited extent to which we can infer the distribution of dark matter in this corona, it appears to be only mildly centrally condensed; a common model is a Plummer-like distribution approximately of the form $\rho(r) \propto \left(r^2 + r_{\text{core}}^2\right)^{-5/2}$ for some core radius r_{core}, which is typically on the order of the radius of the visible disk.

12.3.2 Spheroidal galaxies

In contrast with disk galaxies, the stellar motions in spheroidal galaxies are not organized into a rotationally supported disk but show a mixture of orbital orientations. But they still represent a balance between centripetal and gravitational accelerations as in Equation (12.10). Multiplying each side of that equation by R converts it to

$$2\mathcal{K} = -\mathcal{U} \,,$$

where $\mathcal{K} = \frac{1}{2}V^2$ is the mass-specific kinetic energy and $\mathcal{U} = -GM/R$ is the mass-specific potential energy. Change either one of these energies and either the

[9] Within 100 pc of the midplane; see Section 23.1.10 of *Astrophysical Quantities* (Fourth Edition, A.N. Cox (editor), Springer-Verlag, 2000) for a table of accelerations at differing values of z.

size or shape (or both) of the orbit changes; which is to say, a circular, Keplerian orbit requires this balance between kinetic and potential energies.

The same balance of energies applies as averages in systems with distributed masses and/or non-circular orbits, such as elliptical galaxies and globular clusters. If the system is in stable equilibrium – in practice, neither expanding nor contracting – it must satisfy the **virial theorem**:[10]

$$2\,\langle \mathcal{K} \rangle = -\,\langle \mathcal{U} \rangle \ , \tag{12.13}$$

where (for spheroidal systems) $\langle \sim \rangle$ represents an average over system members at any given time. For spherically symmetric systems with isotropic internal velocities, this equality is realized in terms of observables by

$$3M\sigma_{\mathrm{R}}^2 = \alpha \frac{GM^2}{R} \ , \tag{12.14}$$

where α is a constant of order unity that reflects the internal layering of mass density ($\alpha = 3/5$ for uniform densities), and the factor of 3 on the left-hand side corrects the observed one-dimensional radial velocity dispersion σ_{R}^2 to isotropic, three-dimensional velocities.

In effect, the system radius is that required to accommodate its mass and internal velocities; one infers the system mass from this relation by observing the radial velocity dispersion of system members, and the system radius:

$$M_{\mathrm{virial}} = \frac{3}{\alpha G} R \sigma_{\mathrm{R}}^2 \ . \tag{12.15}$$

As an example: the Fornax dwarf spheroidal galaxy, a relatively large satellite of the Milky Way, shows an internal velocity dispersion of $\left\langle \sigma_{\mathrm{R}}^2 \right\rangle^{1/2} \approx 13$ km/sec. Adopting $R \approx 2$ kpc for this diffuse galaxy yields a virial mass of $\sim 2 \times 10^8$ $\mathrm{M}_\odot$; with an overall luminosity of $\approx 1.5 \times 10^7\ \mathrm{L}_\odot$, the mass/luminosity ratio is $M/L \approx 13$ in solar units. This is much larger than the 2–4 expected from the observed stellar population, indicative of the presence of large amounts of dark matter. This is fairly typical of dwarf spheroidal galaxies, nearly all of which (that are close enough to observe) have virial masses much larger than those implied by their luminosities.

Large elliptical galaxies appear to have internal M/L ratios of 3 to 5 as inferred both from virial analyses and from star counts. But their (typically) large globular cluster populations, external to the galaxy proper, have velocities indicative of much larger M/L ratios, as high as 100 or more. As with disk

[10] A standard theorem of classical mechanics; see any text on stellar or galactic structure for its derivation.

galaxies, large elliptical galaxies appear to have massive dark matter halos or coronas.

12.3.3 Galaxy clusters

Galaxy clusters contain a substantial fraction of the mass in the Universe, most of it in rich clusters containing several hundred large galaxies. A prime example is the Coma cluster with ~ 400 galaxies of mass $10^{10}\ M_{\odot}$ or more, and many smaller galaxies. Such clusters are crowded and constitute deep gravitational potential wells: the mean separation of Coma galaxies is several tens of kpc and a typical internal velocity is nearly 1000 km/s, so cluster members experience scattering interactions on time scales of $\sim 10^8$ years.

Rich clusters are typically infused with a hot, X-ray emitting gas. If the gas is in hydrostatic equilibrium, the density gradient required to confine it to the cluster is

$$\frac{dP}{dR} = -\rho \frac{GM_{(<R)}}{R^2} .$$

Differentiating the ideal gas law, $dP \propto d\,(\rho T)$ so that the mass of the cluster must satisfy

$$M_{(<R)} \propto -\frac{R^2}{\rho} \frac{d\,(\rho T)}{dR} .$$

We can estimate both $\rho\,(R)$ and $T\,(R)$ from spectral properties of the X-ray emission, and thereby infer both the mass of the hot gas and the total mass of the cluster. Typical results for rich clusters are that the X-ray emitting gas must be ~ 5 times the mass of the member galaxies combined, and that the total cluster mass must be ~ 25 times that of all the member galaxies combined in order for the hot intracluster gas to be retained.

12.4 Gravitational lensing

> "QUERY 1. Do not bodies act upon light at a distance, and by their action bend its rays; and is not this action *(cœteris paribus)* strongest at the least distance?"

(Isaac Newton, *Optics*, Part III; 1704). Newton was probably thinking of light as a fluid or as particles subject to (Newtonian) gravitation, so that a ray of light would be deflected in passing by a massive object; and the above query was posed by him as a suggestion for "...further search to be made by others". The correct analysis, of course, is relativistic; and no doubt Newton would have appreciated the outcome in the form of gravitational lensing, and its many applications in modern astronomy. The subject of gravitational lensing, in all its details, is a big one; see, e.g., Section 6.6 of Mo *et al.* (2010) for a thorough and modern review. In this section we summarize the concepts of cosmological utility.

To first order, the gravitational bending of light can usefully be compared with that encountered in classical optics. The metric describing radially symmetric space-time about a central point was first derived by K. Schwarzschild:[11]

$$ds^2 = -\left(1 - \frac{2GM}{c^2r}\right)c^2dt^2 + \left(1 - \frac{2GM}{c^2r}\right)^{-1} dr^2 + r^2d\theta^2 + r^2\sin^2\theta\, d\phi^2 , \tag{12.16}$$

where M is the gravitating mass. In application to light rays, r is the radial distance from M to a point along the transiting ray. Since $ds = 0$ for light, the effective speed of light at radial distance r is

$$c' \equiv \frac{dr}{dt} = c\left(1 - \frac{2GM}{c^2r}\right) .$$

There is thus a 'gravitational index of refraction' given by

$$n \equiv \frac{c}{c'} \approx 1 + \frac{2GM}{c^2r} ,$$

for weak fields $\left(2GM \ll c^2r\right)$. Since the power of an optical lens is proportional to $n - 1$, we would expect the quantity GM/c^2r to play a central role in gravitational lensing.

12.4.1 Strong lensing

By integrating the Schwarzschild metric along the ray path, the gravitational deflection of a light ray passing a distance b from a point source M can formally be shown to be $2GM/c^2b$ on either side of the encounter (in the limit as $b \to \infty$);[12] so the total deflection is

$$\Delta\theta = \frac{4GM}{c^2b} , \tag{12.17}$$

in the small deflection limit. It follows that the image of a point source lying on the lens–observer axis will be a ring (the 'Einstein Ring') whose angular radius is

$$\theta_E = \sqrt{\frac{4GM}{c^2}\frac{d_{SL}}{d_{OS}\,d_{OL}}} , \tag{12.18}$$

[11] The 'Schwarzschild metric', best known for its applications to the space surrounding black holes. The quantity $2GM/c^2$ is the Schwarzschild radius, delimiting the event horizon of non-rotating black holes.

[12] Peter & Uzan (2000), Section 1.6.2.3.

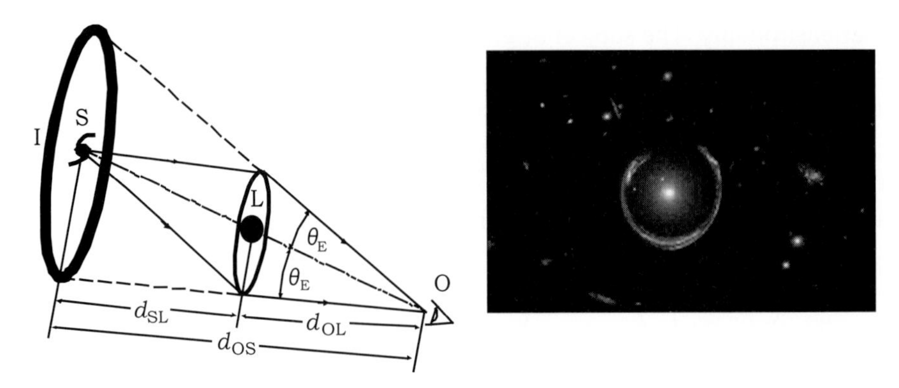

Figure 12.4 On-axis gravitational lensing. Left panel: lensing geometry; the source galaxy (S) is imaged as a ring of angular radius θ_E, the Einstein Radius (angles and off-axis distances greatly exaggerated). Right panel: LRG 3-757, an example of a nearly perfect Einstein Ring (NASA/HST).

where L stands for lens, S for source, I for lensed image, and O for observer. The geometry and an example are shown in Figure 12.4. That the gravitational image of an on-axis point source is a ring, and not a point, is a consequence of the deflection angle being inversely proportional to axial distance from the lens. Just the opposite is the case with ordinary optical lenses where the lens shape is typically such that deflection is nearly proportional to axial distance.[13] As a result, gravitationally lensed images of distant galaxies are badly distorted, although such distortion can often be compensated for in image processing and valuable information about lensed galaxies can be derived.

For off-axis imaging there are two arcuate images straddling the lensed galaxy, whose angular distances from the lens are

$$\theta_{\pm} = \frac{1}{2}\left(\theta_S \pm \sqrt{\theta_S^2 + 4\theta_E^2}\right) , \tag{12.19}$$

where θ_S is the angular separation of source and lens. The geometry is illustrated in Figure 12.5. An example of such gravitational lensing is shown in Figure 12.6: the rich cluster Abell 2218 (comprising most of the fuzzy galaxy images here) has imaged several background galaxies as arcs. From redshifts of these arcuate images, and of the cluster galaxies, the distances in the lensing equation (12.18) can be determined and the mass of the cluster estimated. In this and most similar cases, the cluster masses are compatible with those estimated virially from member velocities and cluster diameter.

[13] A detailed analysis of the relations between gravitational and optical lensing is given by Refsdal and Surdej (1994), *Rep. Prog. Phys.* **56**, pp. 117–185; including designs of oddly shaped optical lenses which mimic gravitational lenses.

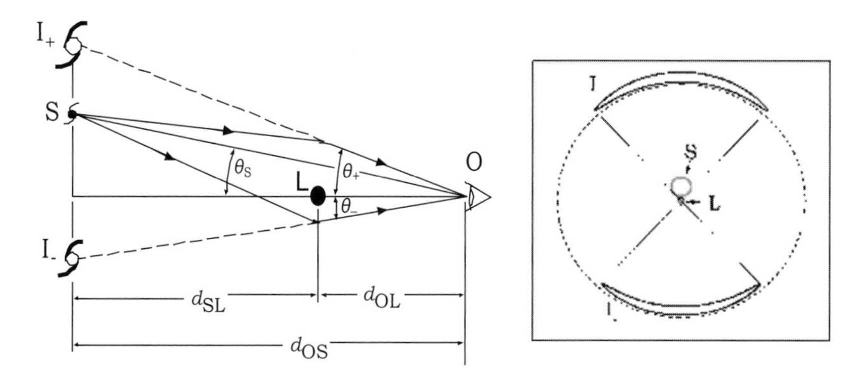

Figure 12.5 Off-axis lensing of a background galaxy (S) by a gravitational lens (L). Left panel: lensing geometry (see text); angles and distances exaggerated for clarity. Right panel: simulated lensing example: the two arcuate images are labelled I_+ and I_-; the dashed circle delimits the Einstein radius.

Figure 12.6 HST image of the rich cluster Abell 2218 and gravitationally lensed images of background galaxies. *NASA*, A. Fruchter and the ERO Team (STScI and ST-ECF).

From Figure 12.6 it is evident that gravitational lensing can produce a magnified image of the source. Since gravitational lensing is a purely geometric effect that does not involve emission or radiation processes, the surface brightness (specific intensity) of an image is unchanged by gravitational lensing, so that the total flux in an image is that of the unlensed image multiplied by the lensing magnification. Gravitational lenses thus can produce optically amplified images of remote galaxies, which are useful tools in the study of distant (hence, young) galaxies.

12.4.2 Weak lensing

The strong gravitational lensing illustrated in these figures produces optical deflections large enough for magnified and brightened images to be well separated from the direct images of the lensed galaxies. Configurations such as these are not common; more often there are no nearly on-axis galaxies to be so imaged by massive clusters or individual galaxies. But galaxies not too far removed from the optical axis might still produce slightly distorted images, whose distortion can be recognized in situations where many lensed galaxies are present. From Figure 12.6 you can appreciate that there is an over-abundance of elongated galaxy images that seem normal, but have their long axis oriented perpendicular to their radial vector from the lensing cluster. Statistical modelling of such orientation biases can produce good estimates of total lensing mass and, in the best of cases, mass distributions within the lensing cluster. This is 'weak lensing': when the lensing cluster is clearly visible, the resulting mass estimates in most cases are once again similar to those estimated with the virial theorem.

12.4.3 Microlensing

Gravitational lensing can arise from lenses of any mass, but only objects with masses in certain ranges can yield lensing images that are observable and useful in inferring lens masses. At the high-mass end are galaxies and galaxy clusters, which produce resolvable, magnified, amplified images as discussed in the previous sections. At the low-mass end are planets and stars (or stellar remnants): their relatively small masses yield Einstein radii that are so small that appreciable lensing requires a very fortuitous arrangement of lens and background object, and whose images are so small as to be unresolvable with current instrumentation. These are **microlenses**, observable only as a brightening of lensed stars caused by the optical amplification attendant upon gravitational magnification of the lensed object's luminosity. Microlenses are most often observed as brightening of background stars by stellar- and planetary-mass objects in the Galaxy's halo or central bulge.

The geometry of microlenses is shown in Figure 12.7. The optical amplification of the lens is[14]

$$\mathcal{A} \equiv \frac{\text{lensed brightness}}{\text{un-lensed brightness}} = \frac{d^2 + 2r_E^2}{d\sqrt{d^2 + 4r_E^2}}, \tag{12.20}$$

where d is the source–lens distance on the plane of the sky at the lens location and

$$r_E = d_{OL}\theta_E = \sqrt{\frac{4GM}{c^2}\frac{d_{OL}\, d_{SL}}{d_{OS}}} \tag{12.21}$$

[14] Schneider (2010); Section 2.5 of this text is a relatively complete review of microlensing physics at an accessible level.

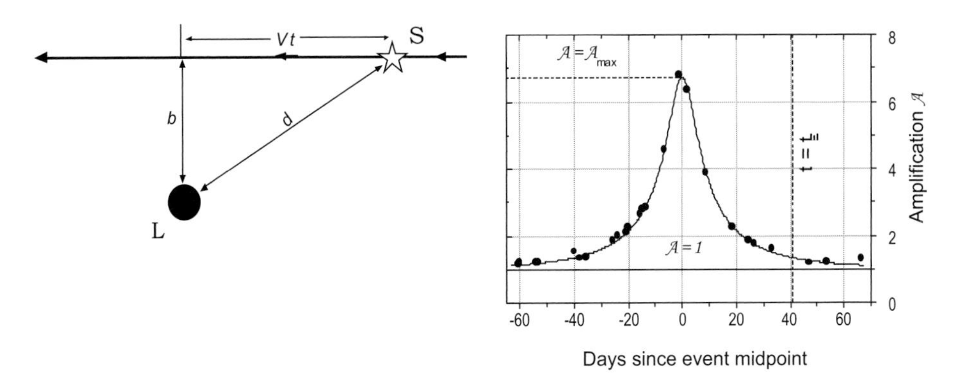

Figure 12.7 Left panel: microlensing geometry; b is the impact parameter, t is computed from event midpoint, and V is the velocity of the source relative to the lens on the plane of the sky at the lens location. Right panel: simulated microlensing event: $b = 0.15\ r_E$, $M = \frac{1}{2}M_\odot$, $d_{OL} = 10$ kpc, $d_{LS} = 40$ kpc, $V = 250$ km/sec.

is the Einstein radius of the lens. The source star moves across the line-of-sight at velocity V (on the plane of the sky) so that its projected distance from the lens at time t (reckoned from the point of closest approach) is $d(t) = \left(b^2 + V^2t^2\right)^{1/2}$, where b is the distance at closest approach. Scaling everything with the Einstein radius,

$$\beta \equiv \frac{b}{r_E}, \quad t_E \equiv \frac{r_E}{|V|}, \tag{12.22}$$

$$\delta(t) \equiv \frac{d}{r_E} = \sqrt{\beta^2 + (t/t_E)^2}, \tag{12.23}$$

in terms of which the amplification may be written as:

$$\mathcal{A}(t) = \frac{\delta(t)^2 + 2}{\delta(t)\sqrt{\delta(t)^2 + 4}}. \tag{12.24}$$

In a microlensing event the source star passes nearly behind the lensing mass so as to produce a light curve, $\mathcal{A}(t)$, similar to that shown in Figure 12.7. By fitting Equations (12.23) and (12.24) to the light curve one can infer values for the scaled impact parameter β and event time scale t_E. In most cases there are no useful constraints on the event impact parameter b and so no information on the Einstein radius can be deduced from the amplification scale of the lensing event; but in many cases the kinematics are sufficiently well known that a likely velocity for the encounter may be estimated and the Einstein radius then inferred as $r_E = |V|t_E$, from which the lens mass may be estimated from known or estimated distances to the lens and source star (Equation (12.21)).

A typical Einstein radius for an observed stellar mass microlens in the Galaxy's halo is on the order of a few astronomical units, so detectable microlensing events are very rare. The best hope for observing them lies with programs that repeatedly image a large number of background stars through the Galaxy's halo, looking for the occasional brightening of a star in the unique (and achromatic) manner illustrated in Figure 12.7. Survey programs of this type have been carried out against the backgrounds of the Magellanic Clouds (MACHO, EROS) and the galactic bulge (OGLE), in the hope of detecting massive dark objects in the halo and in the bulge. The scales of these observing programs are formidable: typically a few million stars are repeatedly observed, from which a large number of light curves must be examined to filter out the occasional microlens. From a cosmological perspective, the most significant conclusion from these surveys is that detectable microlenses are so infrequent that only a small portion of the dark matter resident in the halo can be in the form of microlensing-capable objects with masses ranging from planetary to stellar. The galactic halo dark matter – comprising most of the galaxy's mass – cannot be in the form of Massive Compact Halo Objects (MACHOs), but must be something else.

12.5 Dark matter

The mass density of *all* forms of matter are revealed in the dynamics and lensing observations discussed in the previous two sections. The two types of analysis give results that are largely concordant; in summary,[15]

Mass/luminosity ratios

Large galaxies:	$M/L \approx 10 - 100$,
Galaxy groups & clusters:	$M/L \approx 100 - 500$.

About 10% of all galaxies are in groups or clusters; the remaining are isolated, or nearly so. A reasonable overall mass/luminosity estimate for the current Universe is thus $M/L \approx 200 - 300$ (with very large error bars, and generous allowance for possible undetected matter). Since the luminosity density of the current Universe is observed to be $\mathcal{L} \approx 1.5 \times 10^8$ $L_\odot/\mathrm{Mpc}^3$, the overall mass density of the Universe is probably $\mathcal{M} \approx 3 - 5 \times 10^{10}$ $M_\odot/\mathrm{Mpc}^3$ and the *detected* mass density parameter is probably in the range

$$\Omega_{\mathrm{m},0} \approx 0.25 \pm 0.05, \tag{12.25}$$

[15] Schneider (2010), esp. Fig. 8.11.

This includes all forms of matter, both baryonic and dark. Although substantial amounts of matter may yet be undetected, it seems unlikely that the mass density of the Universe could reach closure values.

Given the small and relatively well-determined density of baryons (Equation (12.9)), the current density of dark matter alone must lie in the approximate range

$$\Omega_{\mathrm{dm},0} = \Omega_{\mathrm{m},0} - \Omega_{\mathrm{baryons},0} \approx 0.20 \pm 0.05, \tag{12.26}$$

or between ~4 and 6 times that of baryonic matter. Dark matter dominates the matter content of the Universe, but is insufficient to close it. As of this writing the nature of dark matter is unknown. The relatively small number of detected halo microlensing events strongly implies that dark matter in our Galaxy cannot be in the form of massive objects, collectively known as Massive Compact Halo Objects, or MACHOs. There is no reason to suppose the situation is different in other galaxies with large amounts of dark matter. The obvious alternative to MACHOs is that of Weakly Interacting Massive Particles, or WIMPs; an acronym apparently chosen to demonstrate that cosmologists are capable of wry humor. These hypothetical creatures are presumed to be heretofore undetected elementary particles, and a great deal of energy is currently being expended in the search for them. The currently favored candidates for WIMPs are particles proposed for *supersymmetry*, a presumed symmetry between fermions and bosons that entails new families of massive particles that have yet to be directly observed; detection of such particles is a major objective of experiments with an upgraded LHC.[16]

It is, of course, always possible that dark matter could reside in forms intermediate in mass between those of elementary particles and planets: the Universe is so large and so nearly empty that there is plenty of room in which to hide dark objects of modest mass, but there is no independent reason to suppose such objects could exist or be formed. It seems possible that the issue of the nature of dark matter will be resolved in the near future, but no prudent person would bet on the outcome.

Our current estimates for mass/energy densities in the Universe – guided by the above arguments and by fits of observations to expansion models as demonstrated in Chapter 15 – are graphically summarized in Figure 12.8. Matter of all sorts apparently comprises about 27% of the Universe's mass/energy budget, the remainder is dark energy (next chapter). Baryonic matter makes up only ~ 4% of the Universe, or ~ 15% of all matter; the remaining 85% is dark matter. Luminous or visible matter makes up only ~ 15% of the baryonic matter, or less than 1% of the Universe's mass/energy budget.

[16] See Section 6.4 of Sanders (2010) for a concise discussion of possible dark matter candidates.

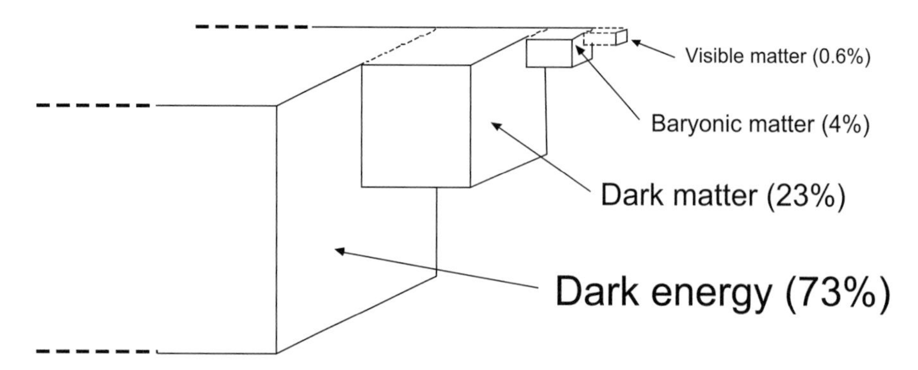

Figure 12.8 The mass/energy budget for the Universe: 73% of the gravitating mass/energy is in the form of dark energy; the remainder is in the forms of matter as indicated.

12.6 Modified Newtonian dynamics

A peculiarity of dark matter is that its presence is not detectable in all astronomical systems, but only in large ones: galaxies and clusters of galaxies, but not in star clusters. The property apparently required for the presumed presence of dark matter is *low acceleration*: dynamical evidence for dark matter is indicated only, and always, in systems with internal accelerations on the order of $a_0 \approx 10^{-10}$ m/s^2 or less. This is the case with, e.g., rotations of spiral galaxies, but not with Solar System orbits. The correlation between evidence for dark matter and gravitational accelerations was first noted by Mordecai Milgrom, who proposed modifications to either Newtonian gravitation or Newton's second law of mechanics, to account for it: Modified Newtonian Dynamics (MOND). Thus, in the simplest case, Newton's law of gravitation would be modified to

$$F = G\frac{Mm}{r^{\alpha}}, \tag{12.27}$$

where $\alpha < 2$ in the low-acceleration regime, where $F/m < a_0$ for baryonic matter. This effectively induces additional 'mass' in conditions of low gravitational acceleration and is quite sufficient to explain nearly all cases of dynamical evidence for dark matter; with suitable relativistic extensions it can also account for evidence from gravitational lensing.[17] But MOND probably cannot account for observed structure in the CMB radiation, and so is not favored by most cosmologists. The correlation between low accelerations and dynamical evidence for dark matter remains unexplained, however.

[17] See chapter 10 of Sanders (2010) for a considered discussion of the evidence supporting MOND as a replacement for dark matter.

Problems

1. Derive an equation for the current age of matter-only models in terms of the current density Ω_0 and Hubble Constant H_0. What is the maximum possible age for a matter-only model with $H_0 = 72$ km/sec/Mpc?
2. From the Milky Way Galaxy's observed rotation velocities discussed in Section 3.2 of this chapter, we expect the Galaxy's dark matter halo to contain $\sim 10^{11}$ $M_\odot$ within ~ 20 kpc of the Galaxy's center.

   (a) Assuming all the dark matter is in the form of MACHOs of mass m, and is roughly uniformly distributed in a sphere of radius 20 kpc, what would be the mean number density of MACHOs, and their mean separation, in terms of m?
   (b) Estimate the mean number of directly detectable MACHOs, assuming (1) a mean MACHO mass comparable to the largest Oort Cloud/Kuiper Belt objects of $m \approx 10^{22}$ kg; and (2) a mean MACHO mass comparable to the smallest detectable asteroids of $m \approx 10^{10}$ kg. Large Oort cloud/Kuiper Belt objects may be visible at distance up to $\sim$100 AU; small asteroids at distances up to $\sim$0.1 AU (1 AU $\approx 1.5 \times 10^{11}$ m). Comment on the delectability of possible MACHO populations.

3. Derive Equation (12.12) for the gravitational acceleration of an exponentially stratified, plane-parallel galactic disk, assumed to extend to infinity.
4. The following table lists data on dwarf spheroidal galactic satellites of our Milky Way Galaxy.[18] Distances (kpc) and velocities (km/sec) are with respect to the Sun. Use these data to estimate the mass of the Milky Way Galaxy. You may assume that satellite galaxies are isotropically distributed about the galactic center. State any assumptions you have made in deriving your mass estimate, and assess its likely credibility.

Galaxy	Distance	Radial velocity
Fornax	140	53
Sculptor	88	107
Leo II	205	76
Sextans	85	225
Draco	80	295
Carina	95	223
UMi	70	247
UMa	100	52
Leo I	270	285

[18] From table 4.1 of Sparke and Gallagher (2007), exclusive of the nearby Magellanic Clouds and Sagittarius dSph.

13

Dark energy

Models containing only matter and radiation cannot account for the observed expansion of the Universe as revealed by (among others) the Hubble Relation for distant supernovae. The discrepancy between models and observations first became clear with the publication of Hubble Relations for type Ia supernovae (SN Ia) in the late 1990s, as demonstrated in Figure 13.1. From this graph it is apparent that the SN Ia data do not fit well to *any* of the matter- (or matter+radiation-) only models. These distant objects appear to be too faint for their redshifts – hence seen at larger distances – than would be expected in standard models. There are several possible explanations for this discrepancy, including mis-calibration of luminosities underlying the relation. But the preferred interpretation is that of a third form of energy density in the Universe, in addition to those of radiation and matter. Rather unfortunately, that component has come to be called **dark energy**, although the *dark* appellation is largely meaningless: *mysterious energy* might be more appropriate, if less evocative. The exact nature of this energy is quite unknown at present, leaving open several possibilities for its form.

13.1 Forms of dark energy

Luminosity distances in a flat Universe are related to the expansion function by Equations (9.43) and (9.83):

$$d_{\mathrm{L}}(z) = (z+1)\,c \int_{(1+z)^{-1}}^{1} \frac{1}{a}\frac{da}{\dot{a}(a)} = (z+1)\,c \int_{(1+z)^{-1}}^{1} \frac{da}{a^2 H(a)} \quad \text{(flat)}\,. \qquad (13.1)$$

If luminosity distances are to be systematically greater – and brightnesses systematically less – in the actual Universe than in the matter-only models, the expansion

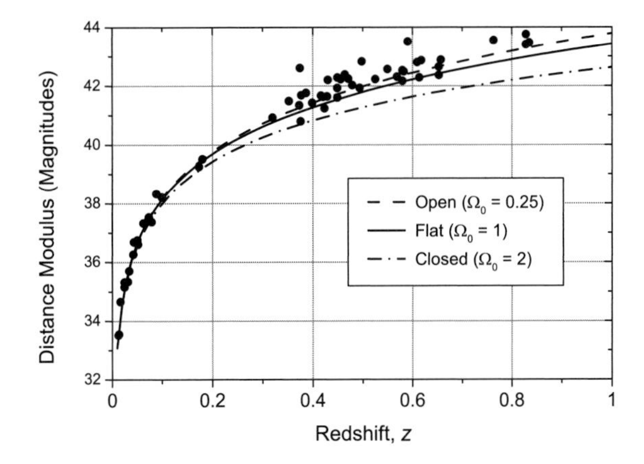

Figure 13.1 Brightness–redshift relations (Equation 9.67) for matter-only models of differing geometries, all for $H_0 = 72$ km/sec/Mpc. The solid dots are the Perlmutter SN Ia data of Figure 14.4.

rate ($\dot{a}$, H) must be systematically less over the range of a incorporating the observations. Of the many mathematical possibilities for such behavior, the physically most plausible seems to be an accelerating expansion driven by the Cosmological Constant. In this picture, $\dot{a}$ would have been less in the past than is the case with standard models, then accelerated to 'catch up to' the current expansion. It was such reasoning that led to the invocation of the Cosmological Constant as a component of the Universe's energy density. The appeal to the Cosmological Constant was bolstered by a suggestive (if tenuous) connection to known physics, in the form of *vacuum energy*.

13.1.1 Vacuum energy

Ordinary quantum fields can create a non-zero energy density in otherwise empty space by creation of virtual particles. Quantum theory predicts that particles of energy ΔE may be created *ex nihilo* and exist for a time Δt satisfying the Heisenberg uncertainty relation:

$$\Delta t \, \Delta E \lesssim \hbar \approx 10^{-34} \text{ J-s} .$$

An electron and its anti-particle (positron) may thus be created and stick around for up to $\sim 6 \times 10^{-22}$ seconds before they annihilate each other. At any time there is thus an equilibrium abundance of such virtual particles that adds up to a 'vacuum energy density' that does not change as the Universe expands. Such a constant energy density is a characteristic of the cosmological constant: $\varepsilon_\Lambda = c^4 \Lambda / (8\pi G)$, so that vacuum energy is a possible source of the Cosmological Constant; and, hence, of dark energy.

The consequence of such a constant energy density $\varepsilon_{\rm de}$ is shown by the resulting Acceleration Equation:

$$\ddot{a} = \frac{H_0^2}{\varepsilon_{\rm c,0}}\left[-\frac{\varepsilon_{\rm r,0}}{a^3} - \frac{\varepsilon_{\rm m,0}}{2a^2} + a\varepsilon_{\rm de}\right].$$

At sufficiently large values of a the acceleration becomes positive, as (apparently) observed. The cosmological issue then becomes one of the magnitude of the resulting energy density or, equivalently, the rate of production of virtual particles. And here there are fundamental difficulties.

In computing likely values for the vacuum energy density one encounters a problem reminiscent of that associated with the 'ultraviolet catastrophe' conundrum preceding the quantum revolution in the late nineteenth century. If all modes of emission (i.e., frequencies) of thermally generated light carry the same amount of energy, as predicted by classical statistical mechanics, the total radiant energy of a hot object should be infinite. Planck saved physics from this apparent paradox by introducing radiation quanta, which incidentally opened the floodgates to quantum physics.

In the case of vacuum energy, standard quantum field theories appear to predict equal contributions to the energy density from all normal modes of the field, corresponding to virtual particle energies (or masses). To avoid an infinite energy density one cuts off the virtual particle production at some maximum mass, $m_{\rm max}$; the resulting energy density is proportional to $m_{\rm max}^4$. But for any plausible maximum particle mass the resulting vacuum energy density is far too large to be reconciled with the observed Universe: for the Planck mass $m_{\rm Planck} \approx 2.2 \times 10^{-8}$ kg, for instance – a natural choice for gravitational fields – the resulting estimate of vacuum energy density is $\varepsilon_{\rm vac} \approx 6 \times 10^{111}$ J/m^3, or about 121 orders of magnitude greater than the $\varepsilon_{\rm de} \sim 6 \times 10^{-10}$ J/m^3 inferred from the Cosmological Constant implied by SN Ia Hubble Relation and other observations.[1] This is a gross mismatch, even in a field (such as astronomy) in which crude estimates are common.

The difficulty probably cannot be resolved by other choices of a maximum mass for virtual particles. If we adopt $m_{\rm max} = m_{\rm nucleon} \approx 1.7 \times 10^{-27}$ kg, for instance – almost surely a lower limit – we still arrive at a ludicrously large vacuum energy density of $\varepsilon_{\rm vac} \approx 1.7 \times 10^{35}$ kg/m^3, about 45 orders of magnitude too large. The vacuum energy density predicted by current quantum field theory is thus larger than what we observe as dark energy by many tens of orders of magnitude and is, in fact, so large that were it to be realized in the early Universe the resulting exponential expansion would have led to nothing like the current

[1] See, e.g., Section 6.3 of Amendola and Tsujikawa (2010) for details of this 'fine tuning problem'.

Universe. Vacuum energy *per se* cannot be the origin of dark energy unless it is offset by some other mechanism to within better than 1 part in 10^{120}; this bit of fine tuning would seem to call for new physics for its explanation.

13.1.2 Quintessence

The generic alternative to vacuum energy would be a previously unknown force of repulsion, archaically and romantically called 'quintessence', that acts only on cosmological scales to modify the expansion rate so as to mimic acceleration.[2] The dynamical nature of quintessence is conveniently characterized in terms of its EOS parameter, which can (in principle) take on any value and is not necessarily constant; but which must lead to universal acceleration if it is to represent a candidate for dark energy. From the second Friedmann Equation (8.20) the effect of this additional energy component on acceleration will be

$$\begin{aligned}\frac{\ddot{a}}{a} &= -\frac{4\pi G}{3c^2}\left(\varepsilon_{\rm r} + \varepsilon_{\rm m} + \varepsilon_{\rm de} + 3P\right) \ , \\ &= -\frac{4\pi G}{3c^2}\left[2\varepsilon_{\rm r} + \varepsilon_{\rm m} + \varepsilon_{\rm de}\left(1 + 3w_{\rm de}\right)\right] \ .\end{aligned}$$

For the acceleration to become positive we must have $\varepsilon_{\rm de}\left(1 + 3w_{\rm de}\right) < 0$, so a necessary condition for acceleration is $w_{\rm de} < -1/3$; a sufficient condition is dependent upon possible variations in $w_{\rm de}$ with expansion.

If $w_{\rm de}$ is a constant,

$$\varepsilon_{\rm de} = \varepsilon_{\rm de,0}\, a^{-3(1+w_{\rm de})} \tag{13.2}$$

and the second Cosmological Field Equation (8.20) may be written as

$$\frac{d^2a}{dt^2} = -\frac{H_0^2}{2\varepsilon_{\rm c,0}}\left[2\frac{\varepsilon_{\rm r,0}}{a^3} + \frac{\varepsilon_{\rm m,0}}{a^2} + \left(1 + 3w_{\rm de}\right)\frac{\varepsilon_{\rm de,0}}{a^{2+3w_{\rm de}}}\right] \ .$$

If $w_{\rm de} < 0$ the dark energy term will eventually come to dominate the expansion, and if $w_{\rm de} < -1/3$ that term will be negative and the expansion will eventually accelerate. In models with a constant dark energy equation of state parameter, $w_{\rm de} < -1/3$ is sufficient and necessary for eventual acceleration. The corresponding expansion equation is

$$\frac{da}{dt} = H_0\left[\frac{\Omega_{\rm r,0}}{a^2} + \frac{\Omega_{\rm m,0}}{a} + \frac{\Omega_{\rm de,0}}{a^{1+3w_{\rm de}}} - \left(\Omega_0 - 1\right)\right]^{1/2} \ . \tag{13.3}$$

The resulting expansion behavior is as follows.

[2] Quintessence, literally, 'fifth essence', originally (*ca.* 1400–1500) referring to a hypothetical fifth fundamental element in addition to Earth, Air, Fire, and Water.

- If $-1 < w_{\rm de} < -1/3$ dark energy density decreases as expansion goes on: $a(t) \to t^x$ with $x > 1$, and the model eventually accelerates at a sub-exponential rate.
- For $w_{\rm de} = -1$ dark energy density is constant and the expansion eventually goes as $\dot{a} \propto a$, characteristic of exponential growth; this is the case for dark energy arising from the Cosmological Constant.
- If $w_{\rm de} < -1$ the energy density *increases* with expansion, which eventually becomes $a(t) \propto \left(t_{\rm rip} - t\right)^{-x}$ with $x > 0$: these models not only accelerate but reach an infinite expansion rate in a finite time $t_{\rm rip}$. Dark energy of this exotic form is commonly called 'phantom energy' and entails a finite lifetime for the expanding Universe.

Our current best estimates for the value of $w_{\rm de}$ – assuming it to be constant – lie roughly in the range from -0.9 to -1.1, so such non-standard possibilities as sub-exponential expansion or phantom energy cannot yet be ruled out.

From Equations (9.43) and (13.2) the proper distance in a dark energy model with constant EOS parameter is

$$d_0(z) = \frac{c}{H_0} \int_{(z+1)^{-1}}^{1} \frac{da}{\sqrt{\Omega_{\rm r,0} + a\,\Omega_{\rm m,0} + a^{(1-3w_{\rm de})}\,\Omega_{\rm de,0} - a^2\left(\Omega_0 - 1\right)}}, \tag{13.4}$$

from which such things as luminosity distance may be computed for models with different forms of (constant EOS) dark energy.

If $w_{\rm de}$ varies with expansion factor a (or, equivalently, z) the energy density will go as (Equation (8.27))

$$\varepsilon_{\rm de}(a) = \varepsilon_{\rm de,0} \exp\left[3 \int_a^1 \frac{1 + w_{\rm de}(\alpha)}{\alpha} d\alpha\right],$$

$$\varepsilon_{\rm de}(z) = \varepsilon_{\rm de,0} \exp\left[3 \int_0^z \frac{1 + w_{\rm de}(z)}{1+z} dz\right].$$

The Expansion Equation then becomes

$$\frac{da}{dt} = H_0 \left[\frac{\Omega_{\rm r,0}}{a^2} + \frac{\Omega_{\rm m,0}}{a} + a^2 \Omega_{\rm de,0} \exp\left(3 \int_a^1 \frac{1 + w_{\rm de}(\alpha)}{\alpha} d\alpha\right) - \left(\Omega_0 - 1\right)\right]^{1/2}, \tag{13.5}$$

and other expressions (proper distance, etc.) follow as with a constant EOS parameter.

The possible forms of $w_{\rm de}(a)$ arise from theories of the quintessence field, for which there are currently no useful constraints. Since the dark energy EOS appears in the Expansion Equation it is possible, in principle, to determine its form or value from the variation of the observed Hubble Relation with redshift.[3] Since Equation (13.5) is an integral relation in $w_{\rm de}$, its solution for $w_{\rm de}$ will require differentiation of the variation of Hubble parameter with redshift. Differentiation of data-based functions is a notoriously noisy process: small errors in the function being differentiated typically lead to large errors in the derivative. Estimation of the form of the dark energy EOS parameter from the Hubble Relation will thus require exceptionally good distance/redshift data if the result is to be credible. It is for this reason that most estimates of $w_{\rm de}$ are made from observations of structure, also difficult since most of the Universe's structure predates the era of dark energy.

13.2 Cosmological Constant models

The most straightforward model for dark energy derives from the Cosmological Constant, Λ. The Expansion Equation for Λ-only models is

$$\frac{da}{dt} = H_0 \left[a^2 \Omega_{\Lambda,0} + 1 - \Omega_{\Lambda,0}\right]^{1/2} . \tag{13.6}$$

For flat geometries these reduce to $\dot{a} = H_0 a$, the differential equation for exponential growth:

$$a(t) = \exp\left[H_0 \left(t - t_0\right)\right] \quad \text{(flat, } \Lambda \text{ only).} \tag{13.7}$$

These models are degenerate in that they have no natural time scale (t_0 is a free parameter, unconstrained by H_0) and are non-zero at all times; i.e., they do not correspond to universes starting their expansion from zero. The Hubble parameter $H(t) = (\dot{a}/a)_t = H_0$ remains constant as the model expands, a characteristic of exponential growth; for the currently estimated $\varepsilon_{\Lambda,0} \approx 6.4 \times 10^{-10}$ J/m^3, $H_0 = 0.063$ Gyr^{-1}. The proper and luminosity distances are

$$d_0(z) = \frac{c}{H_0} z \qquad \text{(flat, } \Lambda \text{ only),} \tag{13.8}$$

$$d_{\rm L}(z) = \frac{c}{H_0} z\,(1+z) \quad \text{(flat, } \Lambda \text{ only).} \tag{13.9}$$

In geometrically open models, $\Omega_{\Lambda,0} < 1$ and the solution to Equation (13.6) is

$$a(t) = \sqrt{\frac{1-\Omega_{\Lambda,0}}{\Omega_{\Lambda,0}}} \sinh\left[\sqrt{\Omega_{\Lambda,0}}\, H_0 \left(t - t_0\right) + \operatorname{arcsinh}\sqrt{\frac{\Omega_{\Lambda,0}}{1-\Omega_{\Lambda,0}}}\right]$$
$$\text{(open, } \Lambda \text{ only).} \tag{13.10}$$

[3] The relevant relations are derived in detail in Section 2.5 of Amendola and Tsujikawa (2010).

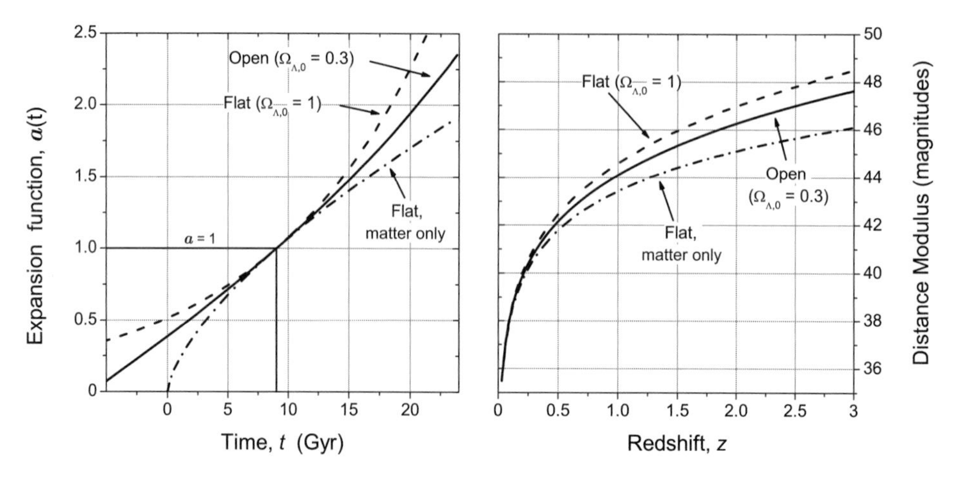

Figure 13.2 Expansion function evolution (left panel) and Hubble relation (right panel) for two Λ-only models (solid and dashed curves); and, for comparison, the flat, matter-only example model (dash-dotted curve); all for $H_0 = 72$ km/sec/Mpc. The time axis labels are for the flat, matter-only model.

The presence of the sinh function implies that this expansion also becomes exponential, $a(t) \propto \exp\left(H_0\sqrt{\Omega_{\Lambda,0}}\,t\right)$, after a few scale times. Current proper distances from Equation (9.44) are

$$d_0(z) = \frac{c}{H_0\sqrt{1-\Omega_{\Lambda,0}}} \ln\left[\frac{\sqrt{1-\Omega_{\Lambda,0}}\,(z+1) + \sqrt{\Omega_{\Lambda,0} + \left(1-\Omega_{\Lambda,0}\right)(z+1)^2}}{1+\sqrt{1-\Omega_{\Lambda,0}}}\right] \tag{13.11}$$

(open, Λ only); from which luminosity distances may be computed with Equation (9.61). Typical forms for Expansion Function and Hubble Relation are shown in Figure 13.2, together with the canonical flat, matter-only model. Note that for given z and H_0, the Λ-model galaxies are further away and fainter than is the case with models not containing a Cosmological Constant. Closed, Λ-only models cannot represent our universe since they do not include dense early phases (as explained in Section 10.2).

13.3 Alternatives to dark energy

13.3.1 Modified Field Equations

The need for invoking dark energy may be circumvented by modifying GR itself. We remind the attentive reader that the Einstein Field Equations (EFE), from which the Friedmann Equations and subsequent expansion functions derive, are not definitive in the same sense as are, e.g., Maxwell's Equations of electromagnetism. The EFE are an inspired guess based in part on analogy to non-relativistic mechanics and, crucially, on the quasi-philosophical preference for simplicity. Several apparently reasonable alternatives to the EFE have been studied, including scalar–tensor theories in which a scalar field couples to the traditional tensor field of GR. Historically, the Brans–Dicke scalar–tensor model (originally proposed in 1961) is of this form and has been extensively studied, as has the Hoyle–Narlikar theory.[4] More exotically, additional spatial dimensions (associated with some modern quantum field theories) have been invoked as a sink for gravitation that would induce acceleration of the Universe. None of these alternative theories have yet proven to be sufficiently compelling as to displace Einstein's original Field Equations as the theory of choice, but neither can they be entirely dismissed.

13.3.2 Inhomogeneous models

Alternatively, the apparent need for dark matter may be a consequence of our adoption of a flawed model for the Universe at large; in particular, of the assumptions underlying the Cosmological Principle. In studying the Universe at large, it is worth keeping in mind that our understanding of the Universe's structure and dynamics largely derives not from actual observations of such things as expansion and acceleration, but from indirect inferences from observations of quantities such as brightness and redshift, based on conceptual models. In particular, the unexpected faintness of SN Ia standard candles is conventionally interpreted in terms of temporal inhomogeneity: the entire Universe was expanding more slowly in past times corresponding to the standard candle redshifts, and hence needed to accelerate in order to 'catch up to' the standard expansion history. But the finite speed of light conflates time and distance, so the SN Ia data may equally well be fitted with a model employing *spatial* inhomogeneity: the distant Universe, in which the standard candles are located, is currently expanding more slowly than is the local Universe. This could obviate the need for acceleration and, hence, for invocation of dark energy.

Consider, for example, two matter-only universes of the same age but differing matter densities. The current expansion rates of the models will vary in an inverse manner with their matter densities.[5] We might thus expect that regions of our Universe of differing densities would currently be expanding at different rates; in particular, that regions more dense than our local Universe would be expanding more slowly. By the reasoning employed at the start of this chapter,

[4] See, e.g., chapter 8 of Narlikar (1993) for discussion of these and other alternative theories.

[5] This is not immediately obvious from the Friedmann Equations, but can readily be verified with numerical experiments with solutions to the equations for coeval models.

standard candles of given redshift in such over-dense regions would be observed at greater distances, and hence be fainter, than those in the local Universe. In effect: if our local Universe is less dense than the remote Universe containing standard candles, the effect on redshifts and brightnesses could be such as to mimic universal acceleration.

There is an obvious flaw in this reasoning: the redshift–brightness relation is based on the RW metric which assumes spatial homogeneity. To investigate the consequences of large-scale density variations we need to develop cosmography based on a metric that supports such inhomogeneities. Probably the simplest such metric, and the one most studied, is the Lemaître–Tolman–Bondi (LTB)[6] metric that explicitly incorporates radially symmetric spatial density variations. The metric replaces RW metric tensor components with ones that are explicit functions of both t and r so that, e.g., $a(t) \rightarrow A(t, r)$. The Einstein Field Equations provide constraints on these metric components in the form of the equivalent of Friedmann relations, but the net effect of the extension of RW metric components to functions of two variables is to so greatly enlarge the solution space as to allow almost any reasonable cosmographical observations to be modelled without invocation of energy densities other than those of radiation and matter. As a consequence, it is possible to model the currently observed Universe in terms of large-scale density variations in which the local Universe is relatively less dense than the distant Universe, and in which the observed redshift–brightness relation of distant standard candles is reproduced without the assistance of dark energy or acceleration.[7] It may not be possible to unambiguously choose between, say, the canonical CCM and such LTB models, from observations of redshifts and brightnesses.

The LTB models have other observational consequences that largely mitigate against their acceptance. But the most cogent objection to such models in most people's eyes is that they do violence to the Copernican and Cosmological Principles. LTB models that successfully describe our currently observed Universe require that we occupy the center of symmetry of a radially inhomogeneous Universe which, while not altogether in conflict with current observations, goes against the grain of 400 years of scientific thought. Perhaps the most useful thing to be learned from such models is that our understanding of our Universe, as embodied in current cosmological models, is only indirectly derived from observations and is subject to interpretations different from those commonly employed.

[6] Named for the three physicists who proposed and studied such metrics in the 1930s and 40s. Chapter 15 of Ellis *et al.* (2012) includes an extensive discussion of the LTB metric and models derived therefrom.

[7] A specific and detailed LTB model that fits current observations without a dark energy component can be found in Enqvist (2008), *Gen. Rel. Grav.* V. **40**, pp. 451–466 (arXiv: 0709-2044 [astro-ph] 2007).

Problems

1. Derive expressions for the event horizons in Λ-only models that are (i) flat, and (ii) open. Use these to show that more open models have larger event horizons (for given Hubble Constant).
2. Compute and graph expansion functions for models containing only matter and dark energy, with $\Omega_{\mathrm{m},0} = 0.27$ and $\Omega_{\mathrm{de},0} = 0.73$, for forms of dark energy with EOS parameters $w_{\mathrm{de}} = -1/2, \ -1, \ -2$. Compare these to the currently accepted CCM expansion model given by Equation (15.2). Comment on the susceptibility of the expansion model to the value of the dark energy EOS.
3. Numerically compute the expansion function for a model in which the dark energy EOS parameter varies with expansion as $w_{\mathrm{de}}(a) = -a$. Using appropriate current energy density parameters, compare this result to those of Problem 2 and comment on the likelihood that w_{de} varies in this manner.
4. Compute and plot expansion functions for two matter-only models differing only in current matter densities and curvature, but of the same age. Verify that the denser of these two models is currently expanding more slowly than the other.

14

Observational constraints

Solutions to the Friedmann Equations constitute a four-parameter family of expansion functions $a(t)$; it is the cosmologists' job to determine which of them is correct, or at least plausible. One way to do so is to measure the Universe's mass/energy densities so as to determine the inputs to the Friedmann Equations, but this approach is inherently inaccurate due to difficulties in determining such densities observationally (Chapter 12) and to the possibility of forms of energy density that leave no direct trace (e.g., dark energy). The alternative is to infer such densities from evidence left by expansion in the form of spatial and kinematic structures in the observable Universe. Most of these hinge upon observational estimates of distances to galaxies.

14.1 Primary expansion diagnostics

Distance estimates on cosmological scales are sufficiently difficult as to have generated a small industry among cosmologists over the past several decades.[1] The overall outline of such methods is hierarchical: distances on one scale are calibrated with those on a previous, smaller scale. Thus:

- a fundamental distance is the Astronomical Unit (AU), conveniently measured by bouncing radar beams off the surface of the Sun;
- distances to relatively nearby stars are then measured in terms of trigonometric parallaxes based on the AU (shifts in apparent star position as the Earth revolves about the Sun);
- distances to other, more distant stars are estimated from stellar properties (e.g., spectral class luminosities) calibrated from nearby stellar parallaxes.

[1] See Section 3.10 of Serjeant (2010) for a concise outline of astronomical distance measurement methods.

All this is sufficient to determine distances within local portions of our Galaxy. To extend this procedure to cosmological distances we require standards that can be seen at such distances and that have either luminosities or other characteristics of known values. Bright objects of known or presumed luminosities are particularly helpful in that they can be used to infer luminosity distances via Equations (9.63) or (9.66); such creatures are known prosaically as 'standard candles'. Historically, the most useful standard candles have been Cepheid variable stars, whose luminosity can be inferred from their readily observable pulsation periods.

14.1.1 Cepheid variables

Classical Cepheid[2] variable stars are horizontal branch (giant) stars with typical masses of $\sim 5 - 10\ M_\odot$ and absolute visual magnitudes ranging from ~ -3 to ~ -6, corresponding to luminosities $\sim 10^3$–10^4 times that of the Sun. Cepheids are pulsationally unstable, the instability arising in helium partial ionization zones in their interiors: they pulsate with periods ranging from ~ 1 to ~ 100 days. More massive Cepheids are more luminous and have longer pulsation periods. The utility of Cepheids stems from the observed correlation of pulsation periods with absolute magnitudes:[3] $\langle M \rangle_V \approx -1.43 - 2.81 \log P$ (P in days), so that their luminosities may be inferred from their observable pulsation periods. This period–luminosity (P-L) relation is the foundation for extragalactic distance estimates but is uncertain at levels of a few tenths of a magnitude. Cepheids are so rare in the solar vicinity that only a handful have reliable trigonometric parallaxes, all measured by the Hipparcos satellite;[4] most of the calibration of the P-L relation is by secondary means, such as membership in galactic open clusters or in the Magellanic Clouds. By virtue of their large luminosities, Cepheids provide the link to extend reliable stellar distance estimates outside our Galaxy: the brightest Cepheids in the nearby Virgo Cluster of galaxies have apparent visual magnitudes $V \sim 25$, observable with modern instruments (Figure 14.1); and have had their periods measured, and thus their luminosities and distances inferred, with Hubble Space Telescope observations.

[2] The name derives from the progenitor star, Delta Cephei, the first such variable studied carefully. The pole star, Polaris, is the brightest (in apparent magnitude) Cepheid variable.

[3] Feast, M. W. & Catchpole, R. M. (1997), *Mon. Not. Royal Astron. Soc.* **286**, p. L1. $\langle M \rangle_V$ is the mean (over the pulsation period) absolute visual magnitude. The P-L relation constants vary somewhat with mean stellar metallicity, hence with color.

[4] Upcoming observations with the Gaia satellite promise improvements in the Cepheid period–luminosity relation, among other aspects of stellar and galactic physics. The progenitor δ Cep also has an apparently accurate parallax from HST observations.

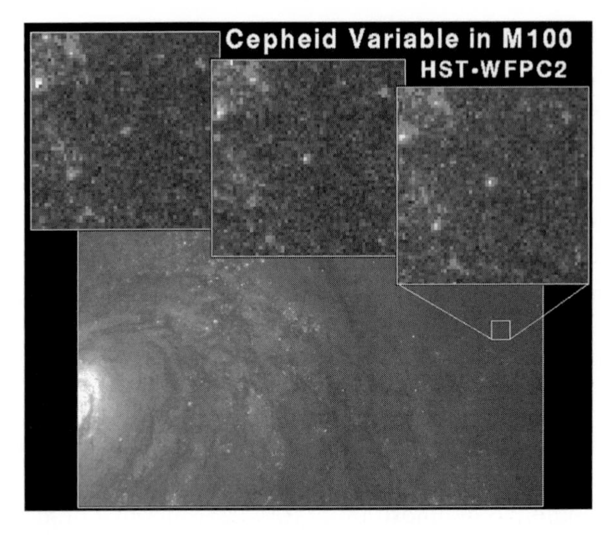

Figure 14.1 Cepheid variable star in the Virgo cluster galaxy M100. The three insets show the star (centered) at three phases in its pulsation period (*NASA*, HST, W. Freedman, R. Kennicutt, J. Mould).

The Virgo Cluster contains a large and varied selection of galaxies,[5] many of them harboring Cepheid variables along with other standard candles not represented in our Galaxy; but luminous and common enough to be found in many external galaxies. Their calibration with Cepheid variables – whose P-L relation has been locally calibrated – enables distance measurements to be extended to cosmological distances. Of the possible standard candles at such distances, the most useful are supernovae of a particular sort.

14.1.2 Supernovae

Exploding stars – novae and supernovae – also have the desirable attribute of high luminosity so that they are visible at great distances. But, taken as a whole, they are poor standard candles, for their luminosities vary greatly. The exception is the Ia class of supernovae, aka 'SN Ia'. Supernovae are classified by their spectral characteristics: SN Ia spectra have no hydrogen or helium lines, show kinematically broadened lines of ionized metals, and have characteristic light curves that decay on time scales of about a month. From observations in nearby galaxies whose distances are calibrated by other means, it is apparent that SN Ia events all reach approximately the same maximum absolute brightness ($M_V \approx -19$ to -20), comparable to that of entire galaxies (see Figure 14.2). Because of their known luminosities, and their visibility at large distances, SN Ia supernovae

[5] About 2000 total, ~ 200 of which are large and luminous, including ~ 140 spiral galaxies more-or-less similar to ours.

Figure 14.2 SN Ia supernova 1994D in the Virgo Cluster galaxy NGC 4526, seen about two weeks before reaching maximum brightness (High-Z Supernova Search Team, HST, *NASA*). Note that the supernova's brightness rivals that of the host galaxy.

have become the *de facto* standard for cosmological distance measurements. SN Ia spectral characteristics are strongly suggestive of the catastrophic explosion of an evolved, degenerate star; one without significant amounts of hydrogen or helium. The preferred model is the 'singly degenerate' one, that of catastrophic collapse of a carbon–oxygen white dwarf resulting from mass transfer from a binary companion.

To appreciate how such a scenario could produce standard luminosities, consider the similar case of a classical nova. Mass transfer from a binary companion builds up on the surface of an orbiting white dwarf; because of the white dwarf's high surface gravity, the hydrogen-rich material deposited onto its surface is heated to the point that hydrogen fusion to helium can take place at its base, blowing off most of the accreted material and brightening the star by (typically) $\sim$ 10 magnitudes. The brightest such novae are visible in nearby galaxies but make poor standard candles because their resulting maximum brightness depends on such *a priori* unknown things as the mass and radius of the white dwarf, and mass transfer rate from the primary; which can probably vary greatly from one nova to another. As a consequence, the absolute magnitudes of novae can vary by more than ten magnitudes amongst the class.

SN Ia progenitors appear to differ from those of novae in that the white dwarf companion is of very nearly the Chandrasekhar limiting mass, so that the accreted material pushes the mass over that limit before the degenerate companion can go nova: the entire star then collapses as a supernova, disassembling the star in a catastrophic explosion. The limited ranges of progenitor masses and of mass transfer rates needed to produce SN Ia explosions, rather than nova-like events, account for the remarkable similarities among SN Ia events. Strictly speaking, SN Ias are not all of exactly the same maximum luminosity, but the variation of their luminosities – over a range of about 1 magnitude – appears to be well-correlated with the (observable) light curve decay time scale, so that all SN Ia light curves can be reduced to the same standard.[6]

It remains somewhat worrisome that the exact nature of these events is not understood: no progenitor object has been observed for any SN Ia, so we cannot be sure that the above described 'singly degenerate' model is correct. A currently popular alternative is the 'doubly degenerate' model, in which two degenerate objects either merge or collide. In any case, SN Ia standardness has been tested in several ways and seems to be reliable.

14.1.3 Dynamic candles

Somewhat surprisingly – given the differing roles of baryonic and dark matter – internal galactic dynamics contain clues as to galaxy luminosities and thus effectively serve as standard candles. The **Tully–Fisher** relation[7] is one between spiral galaxy rotation rates and their luminosities. Since disk galaxy rotation rates are essentially constant outside the central regions (see Figure 12.3), they are well-defined quantities characterizing galaxy disks and measurable by several spectroscopic techniques. Tully and Fisher observed that such rotation rates correlate well with overall galaxy luminosities:

$$L \propto V_{\rm rot}^4 \,. \tag{14.1}$$

Their original work has since been extended to infrared luminosities, in which the relation shows a typical scatter of ~ 0.1 magnitudes in comparison with local galaxies of calibrated luminosities.[8]

A similar relation exists for spheroidal stellar systems, including elliptical galaxies. The **Faber–Jackson** relation[9] is one between system luminosity and its (observable) velocity dispersion σ_0 in the central portions of the galaxy:

$$L \propto \sigma_0^4 \,. \tag{14.2}$$

[6] Phillips, M.M. (1993), *Astrophys. J. Let.* **413**, p. L105.

[7] Tully, R. B. and Fisher, J. R. (1977), *Astron. Astrophys.* **54**, p. 661.

[8] See, e.g., Section 25.2 of Carroll and Ostlie (2007) for further details of the Tully–Fisher relation.

[9] Faber, S. M. and Jackson, R. E. (1976), *Astrophys. J.* **204**, p. 668.

This relation shows a typical scatter of ~ 0.5 magnitudes, which can be considerably reduced by inclusion of one or more additional parameters into the relation. Most commonly chosen are the central surface brightness or the characteristic length scale of the system; the resulting three-parameter relation is commonly referred to as the **Fundamental Plane**.[10]

Both of these dynamical standard candles are useful as secondary distance indicators. But their provenance is worrisome, for they relate a quantity determined by baryonic matter (L) to one largely set by dark matter (internal kinematics). As it happens, MOND (Section 12.6) can readily explain such relations, but 'ordinary' gravitation cannot; at least, not without invocation of some unsupported relations between dynamical quantities within galaxies. The Tully–Fisher and Faber–Jackson relations thus have the current status in standard cosmology of useful but not entirely understood coincidences. At present, this is more a curiosity than a hindrance in determining distances to galaxies.

14.2 Secondary expansion diagnostics

14.2.1 Angular diameters

In principle, the angular diameter–redshift relation (Section 9.3.2) is a sensitive test of expansion models: objects of known physical size ('standard yardsticks') show apparent angular diameters that vary with redshift in a model-dependent manner. But such objects are thin on the ground; in particular, galaxies are evolving structures whose diameters probably increase with age, hence with redshifts; especially during those eras probed by cosmologically interesting redshifts. Perhaps the only really useful cosmological standard yardstick is the sound horizon at the time of photon decoupling, a robust outcome of the Universe's contents and expansion rates, visible as the largest feature in the CMB. Its observed angular diameter is a good match to predictions based on the CCM – see Figure 15.5, and Section 17.3.1 for the details.

14.2.2 Source counts

Again in principle, the variation of galaxy number densities with redshift (Section 9.3.3) is a sensitive test of cosmological models; and again the method is of limited utility because of the likelihood of systematic errors when applied to the observable Universe. Its naive application presumes that the number of *observable* galaxies does not change with time, hence with distance; a very questionable

[10] See, e.g., Section 5.3 of Peacock (1999) for a review of the Fundamental Plane and Faber–Jackson relations.

assumption in brightness-limited surveys (which most are). If only galaxies brighter than a certain limit are detectable, the number of galaxies in the survey will systematically decline with increasing limiting redshift, artificially lowering the values of these curves near the high-z end. In addition, deep surveys may penetrate epochs when galaxies are still forming or evolving to their final luminosities, which would lead to similar observational bias. For these reasons, source count surveys – while very useful in the early days of modern cosmology, when the very notion of universal expansion was controversial and crude source count surveys served to demonstrate its validity – are currently of limited cosmological value.

14.2.3 Surface brightness fluctuations

Stellar brightnesses in external galaxies serve as poor standard candles, but collectively they contain clues to distance in the form of surface brightness fluctuations – by which is meant the pixel-to-pixel variation of brightness in an image of the galaxy. Of two galaxies of otherwise similar stellar content, the more distant will present a smoother appearance. The matter is illustrated in the simulation shown in Figure 14.3.

To first order, the measured surface brightness of a galaxy is independent of its distance: the brightness of a given star falls off with distance as $B \propto d^{-2}$, but the number of stars within a given solid angle – i.e., as imaged onto a detector pixel – varies as $N_{\text{pix}} \propto d^2$, so the total brightness measured by a detector pixel is independent of distance. But the finite number of stars in such images implies a Poisson distribution of stars from pixel to pixel, so the *variation* of pixel brightness with

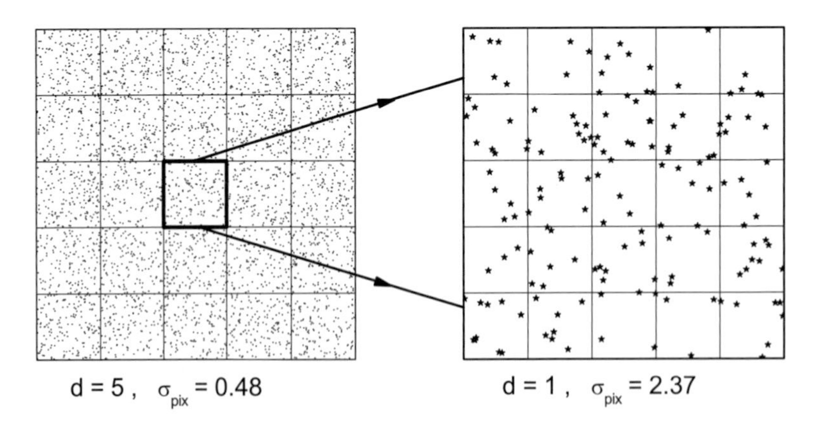

Figure 14.3 Cartoon of simulated images of surface details of a galaxy seen at two distances differing by a factor of 5 (nearest on the right). Grid lines represent pixel boundaries; the central pixel on the left is imaged onto a 5-by-5 pixel array on the right. Pixel-to-pixel brightness variations are larger in the more magnified image (arbitrary units for distances and brightness variances).

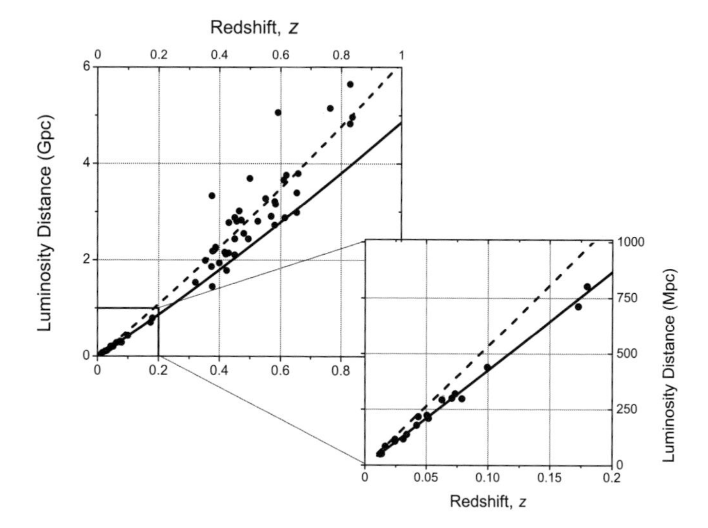

Figure 14.4 Distance–redshift relation for SN Ia standard candles. The two curves are for flat, matter-only models: solid curve, $H_0 = 72$ km/sec/Mpc; dashed curve, $H_0 = 56$ km/sec/Mpc.

distance is approximately $\sigma_{\text{pix}} \propto d^{-1}$. In practice, the application of this technique to infer galaxy distances is most successful for galaxies with smoothly distributed stellar populations: elliptical and S0 galaxies. Even so, there are difficulties with such extraneous matters as variations in brightness distributions within the imaged galaxy due to stellar population variations and interstellar absorption.

14.3 Model validation

14.3.1 Hubble Relation

Standard candles lead directly (via Equation 9.66) to inferred luminosity distances, and hence to observed Hubble Relations; and are relatively free of observational bias. They are thus among the most useful data in determining forms of expansion functions in the Universe or – as is more often the case – rejecting from further consideration those forms yielding Hubble Relations inconsistent with standard candle observations.

An example is shown Figure 14.4: the data are from one of the original SN Ia redshift surveys,[11] the two curves are theoretical ones (Equation (12.3)) for flat, matter-only models of different Hubble Constants. Neither model fits the full range of data: that for $H_0 = 72$ km/sec/Mpc $= 0.074$ Gpc^{-1} (solid curve)

[11] Perlmutter *et al.* (1997), *Bull. Am. Astron. Soc.* **29**, p. 1351.

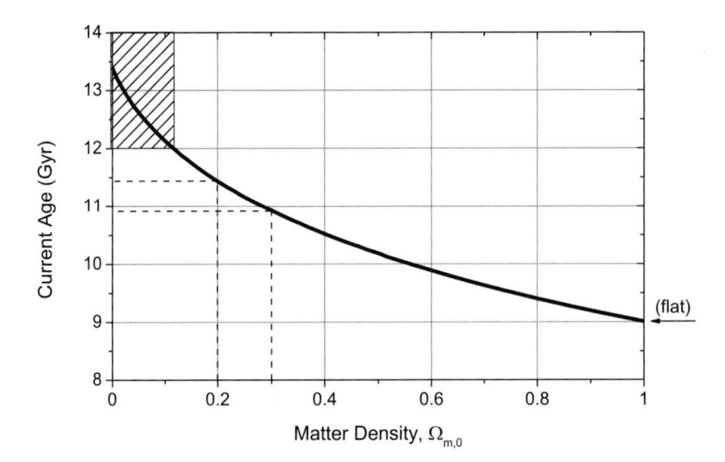

Figure 14.5 Current expansion ages for matter- (or matter + radiation)-only models with $H_0 = 72$ km/sec/Mpc (solid curve). Hatched area: range of ages corresponding to estimated ages of the oldest stars in the Galaxy. Dashed lines: as implied by matter density estimates (0.25 ± 0.05) alone.

is a good fit at small redshifts (inset graph), but a poor fit at higher redshifts; that for $H_0 = 56$ km/sec/Mpc (dashed curve) is just the opposite. This sort of analysis of distance–redshift data suggests that 72 km/sec/Mpc $= 0.074$ Gyr^{-1} is a reasonable choice for the Hubble Constant (since $d_L \approx cz/H_0$ at small redshifts for *all* models; see Equation (9.68)), hence its wide use in this book; but that no flat, matter-only model is consistent with observations of SN Ia standard candles over their entire observed range of redshifts. Extension of this sort of analysis to more extensive and modern data sets (see Figure 15.4) shows that no matter- (or matter + radiation-)only models, including curved ones, are consistent with SN Ia data, Some additional form of energy density appears to be required in order to reconcile the SN Ia observations with plausible cosmological models; as discussed in the previous chapter, dark energy is a likely culprit.

14.3.2 Age

The current age of the Universe is a sensitive function of the model density parameters (see Equation (9.32)); Figure 14.5 demonstrates the relation for models containing only matter.[12] Observationally, the current age can be estimated from several sources,[13] the most reliable and sensitive being the age of the oldest stars in our Galaxy: Population II stars in globular clusters. Age estimates for these

[12] Or matter *and* radiation; the radiation content implied by the CMB is too small to have any significant effect on the Universe's age.

[13] See Section 4.1.3 of Peter & Uzan (2009) for elaboration on age estimates for the Universe.

stars, based on (somewhat uncertain) stellar evolution models, are typically in the range 12 – 14 Gyr, seemingly requiring unrealistically low matter densities for the Universe (hatched region in Figure 14.5). The lowest published estimates for the age of our Galaxy are $\sim$ 11.5 Gyr, marginally compatible with an expansion model containing only matter (dashed lines in the figure), for which the evidence (Section 12.4) is $\Omega_{m,0} \approx 0.25 \pm 0.05$. As with the evidence provided by the SN Ia Hubble Relation, this discrepancy seems to call for additional components in the Universe, or revised physics in our expansion models.

14.3.3 Curvature

Several lines of evidence and reasoning suggest a nearly flat Universe, but the evidence is almost all indirect and inconclusive. Some of it will be shown in Chapter 15, but the best that observational evidence can directly reveal is that the Universe's curvature must be small (in absolute value) in order to produce the observed Hubble Relation and other observational diagnostics. From Section 12.5, $\Omega_0 \gtrsim 0.3$ so that the Universe's radius of curvature satisfies $|R_0| = c/\left[H_0\sqrt{|\Omega_0 - 1|}\right] \gtrsim 5$ Gpc, a proper distance implying $z \gtrsim 2$ in the CCM (Equation (15.6)): rather too early to influence standard candle diagnostics. But the principal motivation for assuming a flat Universe is more one of philosophy rather than of science: since the early Universe was arbitrarily flat, depending upon how early one cares to look (Section 10.1), any non-zero value for the curvature would seem to be arbitrary, so cosmologists generally assume $K = 0$ identically. This may eventually turn out to be wrong, but the possible influence of non-zero curvature on assessment of cosmological models is likely to be very small.

Problems

1. Stellar surface brightness distributions in images of uniformly populated galaxies obey Poisson statistics which, for large numbers, are approximately Gaussian with variances given by $\sigma^2 = \mu$, where μ is the mean number per sample. Using this relation, show that the pixel-to-pixel variation in measured surface brightness of a galaxy is given by

$$\sigma_{\text{pix}} = \frac{\sqrt{n}\Delta\theta}{d} \frac{L_{\text{star}}}{4\pi},$$

where n is the surface number density of stars in the galaxy, $\Delta\theta$ is the angular width of a detector pixel projected onto the sky, L_{star} is the mean stellar luminosity, and d is the distance to the galaxy.

2. The following table lists synthetic brightness and redshift observations for a hypothetical set of standard candles, all of absolute magnitude $M = -18$.

m	z	m	z	m	z
29.12	2.58	24.05	0.41	28.30	1.92
28.92	2.72	25.75	0.78	20.89	0.12
26.38	1.00	29.42	2.86	23.81	0.40
24.48	0.48	28.59	2.09	29.49	2.69
29.14	2.45	27.82	1.58	28.60	2.34
28.06	1.93	26.84	1.59	26.75	1.24
23.27	0.32	27.81	1.63	28.04	1.76
28.50	2.06	28.46	2.16		

Use these data and the Hubble Relation to graphically estimate the Hubble Parameter H_0 and the component energy densities $\Omega_{x,0}$. You may assume that the radiation energy density is insignificant (as revealed by the CMB temperature). Is this universe geometrically open or closed? Is it currently accelerating or decelerating? What is the estimated value for the current matter density, and how does it compare with that of the CCM? Comment on the utility of the Hubble Relation in determining model parameters.

15

Concordance Cosmological Model

The **Concordance Cosmological Model (CCM)**, aka the Standard Model of Big Bang Cosmology, is the solution to the Friedmann Equations that incorporates parameters that represent best estimates taken from a wide range of observations, and that purports to explain not only observations of universal expansion but also consequences of primordial cosmology and of structure formation in the Universe. Perhaps somewhat surprisingly, all these matters can be successfully modelled with one set of parameters for the Friedmann Equations; hence 'Concordance'. The parameters of the CCM are the following.

Curvature From several indirect lines of evidence, the Universe appears to be geometrically flat or nearly so: $\Omega_0 \approx 1$. The principal reason for this choice is the apparent need for a nearly flat geometry in order to account for the Hubble Relation for SN Ia supernovae (Section 15.2), and for the shape of the CMB anisotropy spectrum (Chapter 17). A (very nearly) flat geometry is also a robust prediction of inflation (Chapter 16).

Radiation energy density The well-observed CMB temperature, together with expected primordial neutrinos, implies $\varepsilon_{\mathrm{r},0} = 7.01 \times 10^{-14}$ J/m^3.

Matter density Gravitational evidence (Chapter 12) and model fitting to diagnostics (Chapter 14) point to $\rho_{\mathrm{m},0} \approx 2.6 \times 10^{-27}$ kg/m^3 (corresponding to $\varepsilon_{\mathrm{m},0} \approx 2.4 \times 10^{-10}$ J/m^3), of which $\sim 15\%$ is baryonic (as inferred from primordial nucleosynthesis, Section 16.3); and the remaining exotic dark matter. These densities are also consistent with details in the CMB anisotropy spectrum (Section 17.3).

Dark energy An energy component arising from the cosmological constant of $\varepsilon_{\Lambda,0} \approx 6.4 \times 10^{-10}$ J/m^3 is apparently needed to (1) fit the SN Ia brightnesses to a plausible Hubble Relation, (2) provide enough energy to flatten the Universe, and (3) yield an expansion model in which the Universe is at least as old as the oldest stars in our Galaxy. The nature and density of this energy is uncertain; see Chapter 13.

With the observed $H_0 \approx 2.34 \times 10^{-18}$ sec^{-1}, the critical energy density is $\varepsilon_{c,0} \approx 8.8 \times 10^{-10}$ J/m^3 and the energy density parameters $\Omega_{x,0} = \varepsilon_{x,0}/\varepsilon_{c,0}$ are:

$$\begin{aligned} H_0 &\approx 72 \text{ km/sec/Mpc} = 0.074 \text{ Gyr}^{-1} \Rightarrow \\ \Omega_{\mathrm{r},0} &\approx 8.0 \times 10^{-5} , \\ \Omega_{\mathrm{m},0} &\approx 0.27 , \\ \Omega_{\Lambda,0} &\approx 0.73 , \\ \Omega_0 &\approx 1 . \end{aligned}$$

All of these values seem well established to within several percent. Submillimeter and infrared observations of CMB anisotropies with the recently completed Planck Space Observatory mission are consistent with a somewhat different parameterization: it is not yet clear why the CCM numbers – based on a wide variety of observations, including those with the Wilkinson Microwave Anisotropy Probe (WMAP) – differ from those produced by Planck. The resulting differences in model behavior described below are relatively minor, with the notable exception of current age: $\approx$ 13.8 Gyr with the Planck model vs. $\approx$ 13.4 Gyr with the CCM parameterization.

15.1 Expansion

The Expansion Function for the CCM is numerically computed from Equation (9.21) in the computational form

$$t\,(a) = \frac{1}{H_0} \int_0^a \left[\frac{\Omega_{\mathrm{r},0}}{\alpha^2} + \frac{\Omega_{\mathrm{m},0}}{\alpha} + \alpha^2 \Omega_{\Lambda,0} \right]^{-1/2} d\alpha \, , \tag{15.1}$$

and is shown in Figure 15.1. The accelerating nature of CCM expansion is apparent, as is its relatively older age (in comparison with non-Λ models with the same H_0). Note that for much of the Universe's history up to the present time, the CCM model expands more slowly than do the matter-only models (for the same H_0), but will expand more rapidly in the future.

The onset of acceleration in the CCM is made evident in Figure 15.2. The expansion rate $\dot{a}$ reaches a minimum of $\sim$ 0.062 Gyr^{-1} at $t \approx 6.9$ Gyr (dashed lines) and increases thereafter, reaching the current $\sim$ 0.074 at the current time (solid lines). The Hubble Parameter $H\,(t)$ approaches a constant value (characteristic of exponential expansion) of $\sim$ 60 km/sec/Mpc at large times.

The origins of CCM expansion in gravitating components are shown in Figures 8.2 and 15.3: the expansion is (approximately) composed of successive eras during which each of radiation, matter, and Cosmological Constant dominate the expansion. Noteworthy aspects of the expansion are photon decoupling

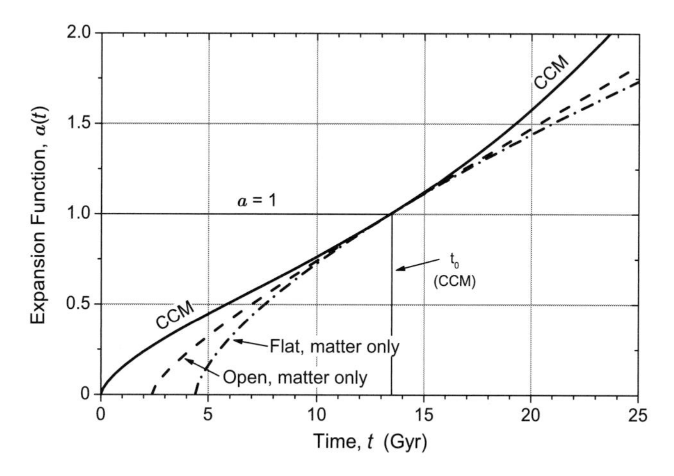

Figure 15.1 CCM expansion function, together with two matter-only models for comparison; all for $H_0 = 72$ km/sec/Mpc. The open model has $\Omega_{m,0} = 0.3$; the time labels refer to the CCM curve.

('recombination') at a time when both radiation and matter densities were substantial, and the switch-over from matter dominance to cosmological constant at very nearly the current time.

Significant mileposts in the expansion are:

Quark–hadron transition*	$t_B \approx 10^{-5}$ sec,	$a \approx 5 \times 10^{-13}$	
Neutrino decoupling	$t_{nd} \approx 1$ sec,	$a \approx 2 \times 10^{-10}$	
$e^- - e^+$ annihilation	$t_{pp} \approx 30$ sec,	$a \approx 5 \times 10^{-10}$	
Primordial nucleosynthesis	$t \approx 100 - 1000$ sec,	$a \approx 2 - 6 \times 10^{-9}$	
Radiation–matter equality	$t_{rme} \approx 52,000$ yr,	$a \approx 3.0 \times 10^{-4}$,	$z \approx 3400$
Photon decoupling	$t_{pd} \approx 375,000$ yr,	$a \approx 9.1 \times 10^{-4}$,	$z \approx 1090$
Acceleration begins	$t_{acc} \approx 6.9$ Gyr,	$a \approx 0.57$,	$z \approx 0.76$
Matter-Λ equality	$t_{m\Lambda e} \approx 9.3$ Gyr,	$a \approx 0.72$,	$z \approx 0.39$
Current time	$t_0 \approx 13.4$ Gyr,	$a_0 = 1$,	$z = 0$

* Prior to quark assembly into hadrons (including baryons) at $t \approx 10^{-5}$ seconds the particle physics are somewhat speculative; after that time they are well understood (excepting dark energy and dark matter).

As is apparent from Figure 15.3, the expansion history of the CCM can be decomposed into two eras, in each of which only two energy components are significant: an early era prior to $a \sim 0.1$ with only radiation and matter, and a following era with only matter and Cosmological Constant. While the expansion function $a(t)$ covering the entire history of the expansion cannot be represented in terms of elementary functions, approximate such formulae may be derived for each of these two-component eras and stitched together for a good approximation to the entire history.

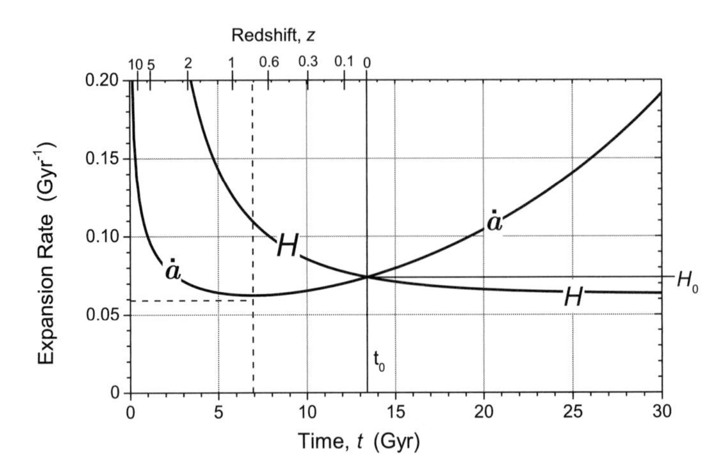

Figure 15.2 Expansion rate evolution in the CCM. Solid lines: current rate $\dot{a} = H_0 = 0.074\ \text{Gyr}^{-1}$ at $t_0 = 13.4$ Gyr. Dashed lines: minimum rate (zero acceleration) $\dot{a} = 0.062\ \text{Gyr}^{-1}$ at $t \approx 6.9$ Gyr, $z \approx 0.76$.

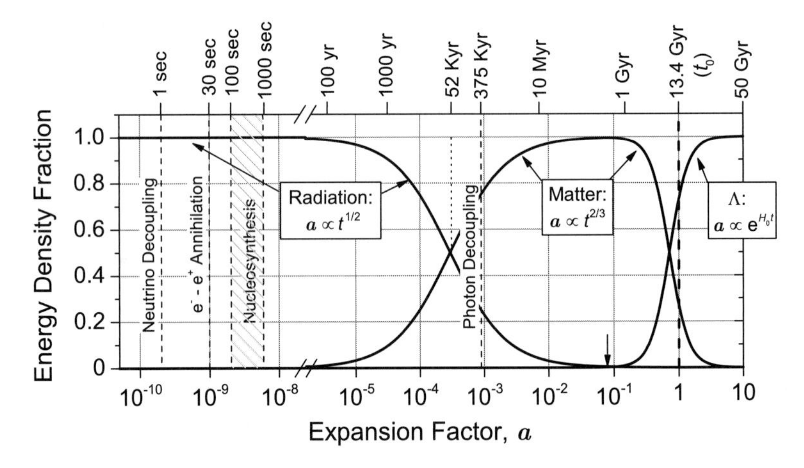

Figure 15.3 Evolution of energy density components as fraction of the total energy, in the CCM.

For a flat model with no radiation an exact solution to the Expansion Equation is

$$a(t) = \left[\sqrt{\frac{\Omega_{\mathrm{m},0}}{\Omega_{\Lambda,0}}}\sinh\left(\frac{3}{2}H_0\sqrt{\Omega_{\Lambda,0}}\,t\right)\right]^{2/3} \quad \text{(flat, no radiation)}\,. \qquad (15.2)$$

This is a good approximation to the later CCM when radiation is insignificant and when galaxies are observable. For the early era in which the cosmological constant is insignificant, an approximation to the expansion is[1]

[1] Adapted from Section 9.3.1 of Ellis *et al.* (2012), which also contains parametric formulae for matter+radiation expansion approximations for non-flat models.

$$t(a) \approx \frac{\zeta(a)^2}{6H_0}\left[\frac{\zeta(a)}{2}\Omega_{\mathrm{m},0} + 3\sqrt{\Omega_{\mathrm{r},0}}\right] \qquad (\text{flat}, \Omega_\Lambda \approx 0), \tag{15.3}$$

where

$$\zeta(a) \equiv \frac{2}{\Omega_{\mathrm{m},0}}\left(\sqrt{\Omega_{\mathrm{r},0} + a\,\Omega_{\mathrm{m},0}} - \sqrt{\Omega_{\mathrm{r},0}}\right). \tag{15.4}$$

Equations (15.2)–(15.4) are excellent approximations to the full CCM expansion function for their respective eras (relative errors less than 1%), for which the switch-over occurs at $t \approx 0.39$ Gyr, $a = 0.08$, $z = 11.5$ (short arrow in Figure 15.3).

15.2 Observational verifications

The CCM predictions for relations among observable variables are largely verified by actual observations. Two of these relations – the Hubble and angular diameter predictions – serve, in effect, to eliminate other simple expansion models from consideration.

Forms of the **Hubble Relations** for the CCM, computed by Equations (9.66) and (9.67), are shown in Figure 15.4, together with corresponding relations for three matter-only models and calibrating SN Ia data. The fit of data to the CCM model is fairly good and clearly better than that for models containing any one component alone, including dark energy (see Figure 13.1). Note that for given

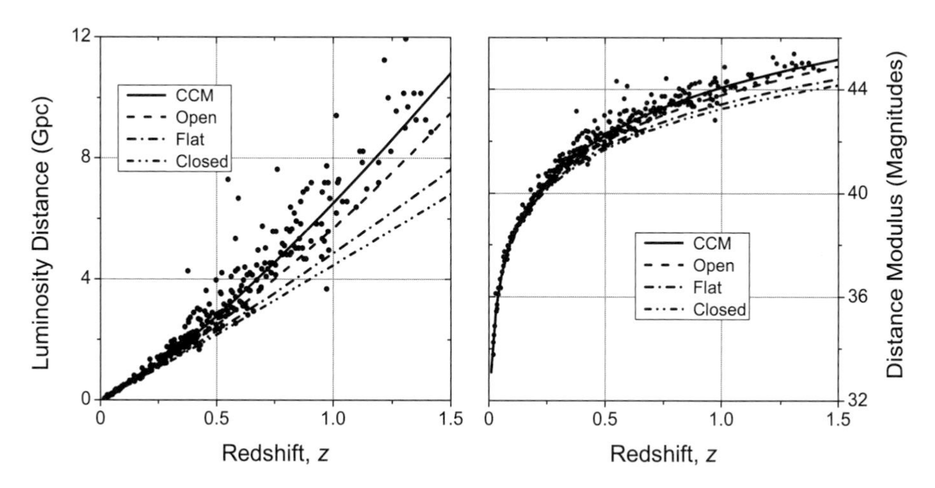

Figure 15.4 Hubble relations for the CCM and three comparison, matter-only models: from top to bottom, $\Omega_{\mathrm{m},0} = 0.3, 1.0, 1.5$; $H_0 = 72$ km/sec/Mpc for all models. Data (solid points) from N. Suzuki *et al.* The Supernovae Cosmology Project (2012), *Astrophys. J.* **746**, p. 85.

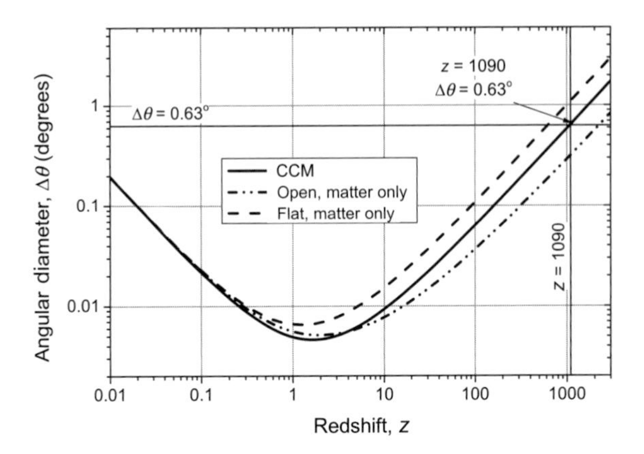

Figure 15.5 Angular diameter vs. distance relation for a $D = 140$ kpc diameter object at various redshifts in the CCM and two comparison, matter-only models with $H_0 = 72$ km/sec/Mpc ($\Omega_0 = 0.27$ for the open model).

redshift, standard candles in the CCM are further away and fainter than is the case with the non-accelerating models.

The apparent **angular sizes** of objects are related to their redshifts by the model-dependent Equation (9.71). Typical such relations are shown in Figure 15.5. The point called out in this figure ($z = 1090$, $\Delta\theta = 0.63°$) corresponds to the sound horizon diameter on the CMB last scattering surface (Section 17.3.1), which is observed to have this angular diameter. The agreement of this datum with the CCM prediction (solid curve) constitutes another corroboration of the CCM.

In principle, the source density relations (Equations (9.75) and (9.76)) for the CCM might also be used to test the model's validity; these are shown in Figure 15.6, for a local source density of $n_0 = 10^8$ per cubic Gpc. But, as argued in Section 14.2.2, observational limitations on surveys of such quantities probably make this technique unreliable in most cases.

15.3 Expansion descriptors

15.3.1 Time

From Equation (9.32) the age of the CCM is about 13.4 ± 0.3 Gyr, gratifyingly close to the estimated ages of the oldest stars in our Galaxy.[2] The coordinate time of emission of a photon arriving from a galaxy of redshift z, and the corresponding look-back time to the galaxy, can be computed from Equations (9.33) and (9.34) for the CCM parameters. The results are shown in Figure 15.7.

[2] The CCM age estimate is sensitive to the adopted value of H_0. Current such estimates range from about 13.2 to 13.8 Gyr.

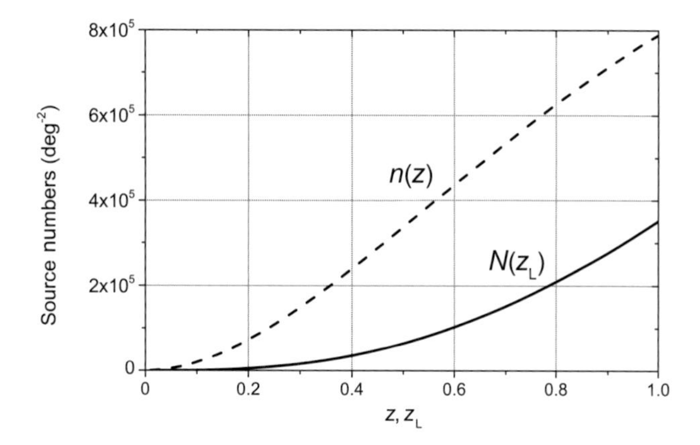

Figure 15.6 Source counts in the CCM with a local source number density of $n_0 = 10^8\ \mathrm{Gpc}^{-3}$. Upper curve: incremental numbers per unit redshift (Equation (9.76)). Lower curve: cumulative numbers out to $z = z_{\mathrm{L}}$ (Equation (9.75)).

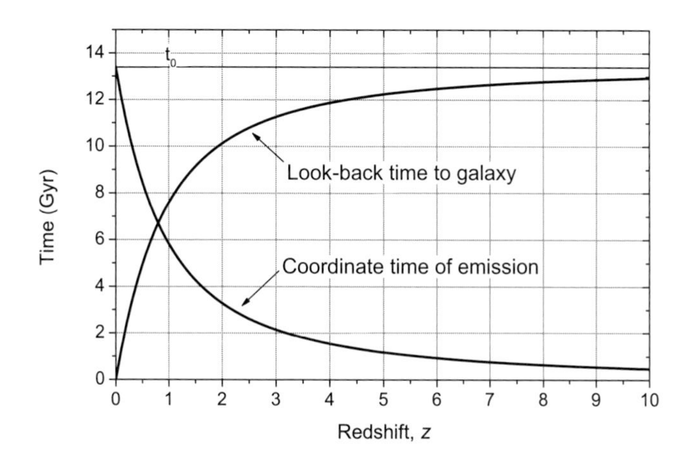

Figure 15.7 Coordinate time of emission $t(z)$ from a galaxy of observed redshift z (Equation (9.33)) in the CCM, and the corresponding look-back time to the galaxy, $t_0 - t(z)$.

Both of these quantities are non-linear functions of redshift. As a consequence, pushing our observations of galaxies to earlier and earlier times in the expansion requires observations of ever-larger redshifts. At present we're routinely (if not easily) observing galaxies of $z \approx 5$, corresponding to an age of $t \approx 1.2$ Gyr; to push that back to $t \approx 0.5$ Gyr when we think galaxies first started forming requires $z \approx 10$. Since the observability of galaxies of the same luminosity goes as $(z+1)^{-4}$ (see Problem 8 of Chapter 9), it will be an order of magnitude more difficult to detect these higher-redshift galaxies.

Two particular times are of interest in this graph. First, our ability to accurately measure cosmological distances rests heavily on one particular standard candle, type Ia supernovae (SN Ia). These have absolute magnitudes $M \approx -19$

and can be detected out to $z \approx 1.5$, corresponding to $t \approx 4$ Gyr. Prior to that coordinate time we must rely on less reliable measures of distance, and our ability to observationally test the predictions of the CCM is greatly reduced. Second, observations of distant galaxies suggest that star formation rates have varied greatly over the history of the Universe, peaking near $z \approx 3$ and tailing off thereafter to a much lower rate in the current Universe. From Figure 15.7 the time of maximum star formation thus corresponds to $t \sim 2-3$ Gyr: star formation was vigorous in the early Universe, more than 11 Gyr before the current time.

For the cosmological era of observational interest – dominated by matter and cosmological constant – the approximate expansion formula (15.2) can be inverted to give an excellent approximation to emission time as a function of redshift:

$$t_{\mathrm{e}}(z) \approx \frac{2}{3H_0\sqrt{\Omega_{\Lambda,0}}} \operatorname{arcsinh}\left[(z+1)^{-3/2}\sqrt{\frac{\Omega_{\Lambda,0}}{\Omega_{\mathrm{m},0}}}\right] \quad \text{(flat, no radiation).} \tag{15.5}$$

This gives CCM time estimates with precisions of 1% or better for $z \lesssim 10$, or $t_{\mathrm{e}} \gtrsim \frac{1}{2}$ Gyr.

15.3.2 Distance

Proper distances are computed from Equation (9.44) with $\Omega_0 = 1$:

$$d_0(z) = \frac{c}{H_0}\int_{(z+1)^{-1}}^{1} \frac{da}{\sqrt{\Omega_{\mathrm{r},0} + a\,\Omega_{\mathrm{m},0} + a^4\,\Omega_{\Lambda,0}}}\,. \tag{15.6}$$

For observable redshifts $\left(z \lesssim 10\right)$ the CCM current proper distance is approximately given by[3]

$$d_0(z) \approx \frac{c}{H_0}\left[\xi\left(0,\Omega_{\mathrm{m},0}\right) - \xi\left(z,\Omega_{\mathrm{m},0}\right)\right], \tag{15.7}$$

where

$$\xi(z,\Omega) = \frac{2}{\sqrt{\Omega}}\left[0.066941\left(\frac{1-\Omega}{\Omega}\right)^{4/3} + 0.19097\left(\frac{1-\Omega}{\Omega}\right)(z+1) + 0.4304\left(\frac{1-\Omega}{\Omega}\right)^{2/3}(z+1)^2 - 0.1540\left(\frac{1-\Omega}{\Omega}\right)^{1/3}(z+1)^3 + (z+1)^4\right]^{-1/8}. \tag{15.8}$$

[3] From Section 1.3.1 of Loeb and Furlanetto (2013).

Since the CCM is flat the corresponding luminosity distance is just

$$d_{\mathrm{L}}(z) = (z+1)\; d_0(z) \; . \tag{15.9}$$

These are excellent approximations to distances in the CCM model and to any other flat model with $\Omega_{\mathrm{r},0} = 0$ and $0.2 \lesssim \Omega_{\mathrm{m},0} < 1$.

15.4 Horizons

Both particle and event horizons, as well as proper distances to galaxies, grow monotonically with time in the CCM. Those galaxies currently in causal contact with us have proper distances less than the particle horizon; those whose current state will become (or remain) visible to us in the future have proper distances less than the event horizon. Configurations of horizons and proper distances in the CCM are most easily seen in terms of co-moving distances which are, in effect, proper distances with the expansion parameter $a(t)$ factored out. Co-moving distances to galaxies are the same as current proper distances and thus constant in time; those of horizons vary as horizons evolve within the co-moving metric structure. The situation is illustrated in Figure 15.8, in which co-moving galaxy distances are represented by horizontal lines (that for $d_0 = 3$ Gpc is shown).

Galaxies currently closer than 4.7 Gpc are within our particle and event horizons and both causally connected and visible to us. Galaxies between 4.7 Gpc (current event horizon distance, lower solid circle) and 14 Gpc (current particle horizon distance, upper solid circle) have been in causal contact in the past, but

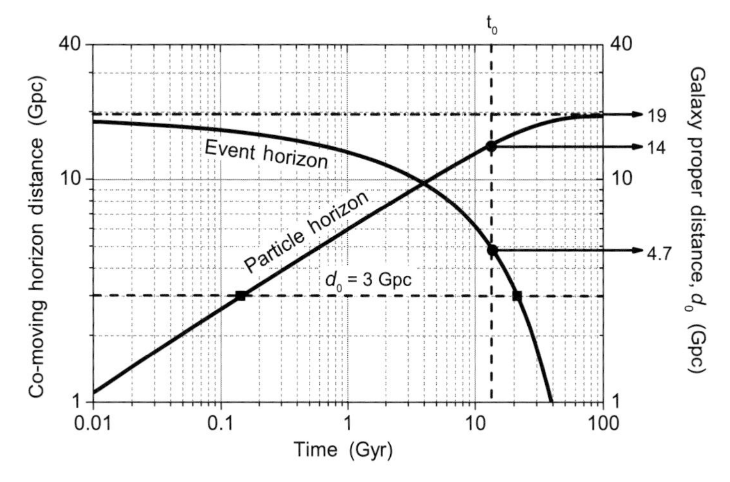

Figure 15.8 Evolution of particle and event horizons in the CCM, plotted as co-moving distances $d_{\mathrm{p}}(t)/a(t)$. Galaxies have constant distances d_0 in this system, denoted as ordinates on the vertical axis. Regions bounded by the horizon curves delineate forms of causal contacts with our Galaxy, over the course of the Universe's evolution; see text.

no longer are so. They are analogous to ex-presidents whose past actions may still affect us, but whose current or future actions cannot (or so we hope). If a galaxy were only now to come into being in this distance interval we would never see it, nor be influenced by it: the Universe is expanding too rapidly.

Galaxies between 14 and 19 Gpc from us have never been in causal contact with us, and will only come into contact in terms of past influences: we will never see them as they currently are. Galaxies with current proper distances greater than $d_{EH}(t=0) = d_{PH}(t \to \infty) \approx 19$ Gpc (dashed-dotted line) have never been in causal contact, and never will be: this distance effectively delimits the size of our knowable Universe.

The dotted horizontal line represents a typical, relatively close galaxy ($d_0 = 3$ Gpc) that enters the particle horizon and thus becomes causally connected to us at about 0.15 Gyr (solid square); it will leave the event horizon and thus lose contact with us at about 20 Gyr (second solid square).

This somewhat complicated arrangement of horizons in the CCM is a consequence of the accelerated expansion arising from the Cosmological Constant. Without such acceleration the particle horizon would grow monotonically with time, and the event horizon would be infinite at all times; so all galaxies would eventually come into and remain within the particle and event horizons, hence in causal contact with us. But the Cosmological Constant causes the particle horizon to stop growing and the event horizon co-moving distance to decrease as the Universe expands, so that our visible horizon contains fewer and fewer visible (and co-moving) galaxies as time goes on. It is in this sense that universal acceleration can be said to shrink the visible Universe over time.

From Equations (10.9) and (10.12) applied to $a(t)$ for the CCM, the current horizon distances in the CCM are

Particle horizon*	$d_{PH} \approx 14.0$ Gpc,	$z \to \infty$
Event horizon	$d_{EH} = 4.7$ Gpc,	$z \approx 1.7$
Visible horizon	$d_{VH} = 13.8$ Gpc,	$z \approx 1090$.

* Note that the particle horizon quoted here (and plotted in Figure 15.8) is computed without inclusion of a possible early inflationary epoch, which would greatly increase the current horizon distance (see Chapter 16). The visible horizon is computed as $d_p(t_{pd})$, Equation (9.42), for $t_{pd} = 3.75 \times 10^{-4}$ Gyr, the approximate time of photon decoupling and CMB formation.

Problems

1. Compute the total baryon mass within the particle horizon of the CCM. If most of this is in the form of galaxies of typical mass $M \sim 10^{11}\ M_{\odot}$, how many galaxies are within our horizon? This is effectively the galaxy content of our observable Universe.

2. The **Steady State** cosmological model arises from the 'perfect cosmological principle' that the Universe is not only homogeneous and isotropic, but is also unchanging in time. This requires that all observable aspects of the Universe are unchanging – in particular, that the Hubble Parameter, energy density, and curvature are all constants; so that ongoing mass creation is needed to keep the density of the expanding Universe constant. The theory was a serious alternative to the Big Bang until evidence for universal evolution, including the CMB radiation, was discovered in the 1960s. It is now more an historical curiosity than a realistic theory.

   (a) Solve the Expansion Equation for $a(t)$ for a steady state model of Hubble Constant H, normalized to $a(t_0) = 1$.
   (b) Derive an expression for luminosity distance for this model and compare the resulting Hubble Relation to that of the CCM.
   (c) If $H = 72$ km/sec/Mpc $= 0.074$ Gyr^{-1}, estimate the rate of matter/energy creation needed to keep the Universe's energy density constant. Would this energy creation be easily observable?

Part V

Expansion history

All credible cosmological models are governed by decreasing energy densities and temperatures as the Universe expands. The very early Universe almost certainly was hot and dense, and probably nearly uniform as a consequence of the high rate of particle interactions likely to have prevailed in that state; and thus largely without structure. The current Universe is quite different, with matter coarsely distributed and large temperature gradients on many scales, and coherent structures of all sorts – from galaxy clusters and superclusters to individual galaxies of all sizes; to planetary systems and astronomy students. How the Universe changed so greatly in $\sim$ 13 billion years is the subject of much study amongst cosmologists, who are also interested in what the progress of those changes can tell us about the Universe at large.

It is useful and customary to break the Universe's history into a series of eras characterized by processes and contents. The selection of eras is largely arbitrary – it constitutes something of a Rorschach test for theoretical cosmologists. For present purposes we choose the following:

- a **Particle Era**, including the earliest times and the Inflation event; the era ends rather arbitrarily with primordial nucleosynthesis of helium and other light elements;
- a **Plasma Era** from the end of the nucleosynthesis epoch until recombination, when free electrons combined onto positive nuclei and photons decoupled from matter; followed by
- the **Galaxy Era** during which galaxies formed and evolved, up to the current time.

Our knowledge of the events in these eras varies considerably. We are particularly ignorant of the details in the early parts of the Particle Era when energies greatly exceed those achievable in particle accelerators, rendering the physics uncertain; and in much of the Galaxy Era where the physics are relatively complicated and observations can be quite difficult. Despite these difficulties, cosmologists have assembled a coherent and apparently reasonable, if certainly incomplete, picture of the Universe's evolution up to the current time.

16

Particle Era

All plausible cosmological models include an early phase that was very hot and dense. The evidence for such a beginning is manifold, including fossil relics such as light elements and the CMB radiation. What is not clear is the nature of the very beginning of the Universe, for our incomplete knowledge of particle and field physics at high energies does not allow us to push theory beyond a certain point in the past.

The early Universe was a hot, dense mélange of elementary particles and radiation that – with the exception of the inflationary epoch discussed in the following section – steadily cooled and became thinner as the Universe expanded. In such a state particle interactions amongst elementary particles are expected to be so frequent that everything is effectively in thermodynamic equilibrium at any given time, irrespective of expansion. In analyzing such a state it is useful to employ mean particle energy as a proxy for the expansion function: since $T(a) = 2.725/a$ K (Equation (11.1)) for the CMB with which everything is in thermodynamic equilibrium, and $\langle \mathcal{E}(T) \rangle = 2.701 k_B T$ in black-body radiation,

$$\langle \mathcal{E}(a) \rangle \approx 6.34 \times 10^{-10} a^{-1} \quad \text{MeV}. \tag{16.1}$$

The Particle Era takes place under radiation dominance of the expansion, so that $a(t) \approx 2.05 \times 10^{-10}\, t^{1/2}$ for t in seconds (Equation (11.8)); and thus

$$\langle \mathcal{E}(t) \rangle \approx 3.1\, t^{-1/2} \quad \text{MeV}, \tag{16.2}$$

for t in seconds. This holds for all the Particle Era following inflation, and for much of the Plasma Era.

But this is only part of the story, for particle interactions can be sources or sinks of energy. As canonical examples, the positron annihilation reaction $e^- + e^+ \longrightarrow \gamma + \gamma$ injects ~ 1 MeV (twice the electron rest energy) per anti-electron into the plasma; and the sequestration of quarks into hadrons decreases the number of relativistic degrees of freedom, hence increasing the energy per particle. A close accounting of such variations is needed for a detailed description of the evolution of the early Universe, but fortunately will not be necessary for our purposes since their effects on mean particle energies are not important to the big picture, and we can safely employ Equation (16.2) as a reasonable approximation to particle energies in tracking the early Universe's evolution. Consult Peter and Uzan (2009), especially Section 4.2, for the details.

Prior to $t \sim 10^{-10}$ seconds, mean particle energies were several hundred GeV and exceeded what is achievable with particle accelerators, so our understanding of particle and quantum field physics must be speculative. With this as a caveat, the evolution of the early Universe is believed to have proceeded as follows.

- At very early times, such as the Plank time ($\sim 10^{-44}$ seconds), energies were so high that all four fundamental forces were united into one. Sometime between then and the GUT transition to follow, gravitation separated from the other three forces in the TOE (Theory of Everything) transition. We do not currently have an acceptable theory of quantum gravitation, and so can only speculate on the timing and nature of this transition.
- At $t \sim 10^{-35}$ seconds, $\langle \mathcal{E} \rangle \sim 10^{15}$ GeV and the strong nuclear force separated from the electroweak force in the GUT (Grand Unified Theory) transition.
- At approximately the same time as the GUT transition, the Universe experienced a brief period of hyper-expansion known as *Inflation.*
- Some time after Inflation, and probably prior to the electroweak transition, an imbalance arose of matter over anti-matter and a net non-zero baryon number density was established by processes collectively known as *Baryogenesis*.
- At $t \sim 10^{-10}$ seconds, $\langle \mathcal{E} \rangle \sim 300$ GeV and the weak nuclear force separated from the electromagnetic force in the electroweak transition; thereafter, all four fundamental forces assumed the forms they have today.
- At $t \sim 10^{-4}$ seconds, $\langle \mathcal{E} \rangle \sim 300$ MeV and quarks assembled into hadrons: protons and neutrons appeared on the scene for the first time. After this time, our knowledge of the early physics of expansion is relatively secure (excepting dark matter and dark energy).
- At $t \approx 2$ seconds, $\langle \mathcal{E} \rangle \sim 2$ MeV and neutrinos decoupled from matter, breaking the thermal equilibrium between protons and neutrons.
- At $t \approx 30$ seconds, $\langle \mathcal{E} \rangle$ fell below the electron rest energy of $\sim 1/2$ MeV and $e^- - e^+$ pair production ceased, immediately followed by positron annihilation reheating the Universe and leaving a fossil remnant of electrons.
- At $t \sim 100$ seconds, $\langle \mathcal{E} \rangle \sim 0.3$ MeV and primordial nucleosynthesis began in earnest. By $t \sim 1000$ seconds it had largely stopped, leaving as fossil remnants the primordial light element abundances we see (or try to see) today.

Following the end of nucleosynthesis, the Universe's contents consisted of a plasma of mostly hydrogen and helium nuclei, and a very small amount of somewhat heavier nuclei; expanding and cooling. This is the Plasma Era, described in detail in the next chapter.

16.1 Inflation

The current cosmological paradigm is one of Λ-dark energy and of *Inflation*, by which is meant a very early, very rapid expansion of the Universe by many orders of magnitude. The initial motivation for a theory of Inflation came from consideration of the consequences of fundamental force separations in the early, high-energy Universe; particularly of the GUT transition which was the highest-energy such event for which some understanding might come from current physics models. Such transitions are consequences of spontaneous symmetry breaking in the underlying fields that can, in some cases, lead to a rapid and exponential expansion. Since symmetry breaking would probably induce topological defects such as magnetic monopoles, and no such defects are observed in the current Universe, the inflationary expansion serves as a welcome *deus ex machina* to sweep the missing particles beyond current horizons, thus accounting neatly for their absence.

Such speculations were in the air in the early 1980s when Alan Guth (1981) noticed that inflation would also help with some pesky problems in modern cosmology. In particular, the observed Universe seemed to be nearly uniform on scales greatly exceeding the horizon distance predicted by then-accepted expansion models, and the available dynamical evidence strongly suggested a nearly flat geometry to the Universe and thus an age less than that of the oldest stars in our Galaxy. Guth observed that an early period of hyper-expansion would both flatten the Universe and greatly expand its horizons, in addition to effectively eliminating monopoles. It is this serendipitous combination of particle physics with cosmology that accounts for the rapid acceptance of the theory of Inflation and of its continued status as a cornerstone of modern cosmology, although current theories of the nature of Inflation are somewhat different from Guth's original proposal.

16.1.1 Inflation mechanism

For reasons deeply embedded in the Standard Model of fundamental particles and fields, the field responsible for Inflation – the *inflaton* field (not a misprint!) – is presumed to be a scalar field of (nearly) constant energy density, quite similar to that of the Cosmological Constant. The dynamical result would be the same as with Λ: accelerating, exponential expansion. If Inflation is to sufficiently flatten the Universe, increase its horizons, and dilute the monopole density to near zero, the expansion extent will probably need to be of a factor of at least[1] $\sim 10^{30}$ in terms of a, so speculative scenarios typically employ expansion factors e^N with N ranging from ~ 70 to ~ 120. Exponential expansion entails a constant Hubble Parameter (Section 13.2), so if exponential inflation starts at t_{is} and goes on until t_{ie} the expansion rate H_I must satisfy $\exp\left[H_I\left(t_{ie}-t_{is}\right)\right] = \exp N$, or $H_I = N/\left(t_{ie}-t_{is}\right)$. The expansion function during Inflation is thus

$$a_I(t) = a_{is}\exp\left[H_I\left(t-t_{is}\right)\right] = a_{is}\exp\left[N\frac{t-t_{is}}{t_{ie}-t_{is}}\right] \qquad \text{(Inflation)}, \qquad (16.3)$$

where a_{is} is the expansion factor at the start of inflation.

As a canonical example to be used throughout this chapter: if we adopt a minimalist expansion extent of $N = 70$ (or ~ 30 orders of magnitude) starting at the presumed GUT transition time $t_{is} = 10^{-36}$ sec and ending at $t_{ie} = 10^{-34}$ sec, we would require an expansion rate of $H_I \approx 7\times10^{35}\ \text{sec}^{-1}$. If $a_{is} = a\left(t_{is}\right)$ is the expansion factor at the start of inflation and $a_{ie} = a\left(t_{ie}\right)$ that at the end, $a_{ie}/a_{is} = \exp N \approx 10^{30}$; all this in $\sim 10^{-34}$ seconds. Rapid expansion, indeed. Since 'ordinary' CCM expansion presumably takes over at the end of inflation it must be that $a_{ie} = a_{ccm}\left(t_{ie}\right) \sim 10^{-27}$, where a_{ccm} is the CCM expansion function computed in Chapter 15 and given by Equation (11.8) at such early times. The relations of pre-inflation, inflation, and post-inflation expansion scenarios are illustrated in Figure 16.1, which restricts the inflation to $N \approx 3$ in order to illustrate the connections between expansion eras.

Since the mean particle energy (or temperature) varies with expansion factor a as $\langle\mathcal{E}\rangle \propto 1/a$, and inflation increases a by a factor of $\exp(N)$ or about 10^{30} in our canonical example, particle energies and temperatures would be greatly reduced by inflationary expansion so that, e.g., we might expect the current CMB radiation temperature to be many orders of magnitude less than what is observed. The Universe is saved from this fate by a brief period of *reheating* at the end of Inflation: the collapsing inflaton field invests its residual energy in creating the particles and radiation that occupy the Universe thereafter. The expanding inflaton field behaves similarly to a super-cooled liquid, retaining its liquid phase down to temperatures well below the normal freezing point, then releasing the pent-up latent heat in a sudden phase transition.[2] In this scenario, all the Universe's observed contents were created from the inflaton field's energy density at the end of Inflation, while those particles in the Universe prior to Inflation – such as magnetic monopoles – have had their densities so diluted as to be effectively removed from the observable Universe.

[1] Justifications for adopting such a large value of N may be found in §4.1 of Weinberg (2008).

[2] A well-observed phenomenon in ordinary water where, with care to keep the water pure and to cool it slowly, the liquid can be cooled to temperatures 20 C below freezing before it suddenly freezes throughout.

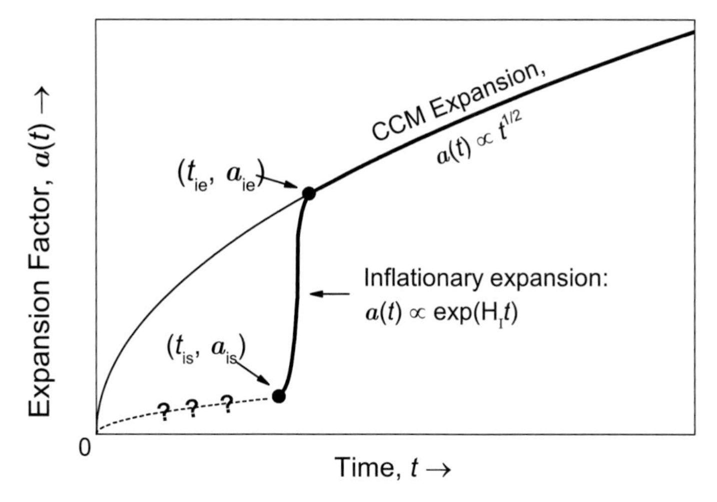

Figure 16.1 Cartoon of universal expansion $a\,(t)$ incorporating Inflation and melding with the CCM expansion (heavy curve). The light, solid curve is the CCM expansion without inflation; the dotted curve with question marks is a speculative $a\,(t) \propto t^{1/2}$ expansion prior to onset of inflation.

16.1.2 Cosmological consequences

Horizons

From Equation (10.9) the particle horizon d_{PH} at the time t_{pd} of photon decoupling and CMB formation was

$$d_{\mathrm{PH}} = c\,a\left(t_{\mathrm{pd}}\right) \int_0^{t_{\mathrm{pd}}} \frac{dt}{a\,(t)} \approx 260 \text{ kpc}\,,$$

according to the CCM expansion function $a\,(t)$. From Equation (9.85) the corresponding horizon angular diameter would be

$$\Delta\theta_{\mathrm{PH}} \approx \left(1 + z_{\mathrm{pd}}\right) \frac{d_{\mathrm{PH}}}{d_0\left(z_{\mathrm{pd}}\right)}$$

for $|\Omega_0 - 1| \ll 1$. For $z_{\mathrm{pd}} \approx 1000$, as is predicted by most models, this amounts to $\Delta\theta_{\mathrm{PH}} \sim 1$ degree and, indeed, the largest structures seen in the CMB are nearly this large. The irregularities ('anisotropies') seen in the CMB thus appear to be within their particle horizons: nothing much larger should be causally connected at that time. There are thus $\gtrsim 10^4$ such horizon patches covering the 4π steradians of the sky, each of them causally disconnected from the rest. It is then surprising that these CMB irregularities (observed as temperature fluctuations) are all so nearly of the same amplitude: the typical relative difference between any two of them is $|\Delta T/T| \lesssim 10^{-5}$. The CMB is remarkably uniform for something composed of many causally disconnected regions.

The same worrisome issue persists into the current Universe. The current particle horizon distance is very nearly the same as the visibility horizon (proper distance to the surface of last scattering that appears to us as the CMB): both are about 14 Gpc. Distant galaxies on opposite sides of the sky are thus separated by more than a horizon distance, but are essentially similar (on average) in all respects. Something thus appears to have 'homogenized' the Universe fairly early in its history. Of course, it is always possible that the Big Bang created a thoroughly uniform Universe *ex nihilo,* but it would be more satisfactory if a mechanism could be found that would greatly expand the Universe's horizons and thus explain the uniformity of the CMB, without appeal to special initial conditions.

Inflation accounts for these horizon issues not so much by making the Universe bigger, but by having it initially be smaller so that horizons could encompass more of it. The growth of particle horizons is governed by Equation (10.9):

$$d_{\mathrm{PH}}(t) = c\,a(t)\int_0^t \frac{d\tau}{a(\tau)}\,.$$

For a non-inflating Universe (thin solid curve in Figure 16.1) at these early times, $a(t) = a_{\mathrm{ie}}(t/t_{\mathrm{ie}})^{1/2}$ so the particle horizon at time t_{ie} is

$$d_{\mathrm{PH}}(t_{\mathrm{ie}}) = c a_{\mathrm{ie}}\int_0^{t_{\mathrm{ie}}} \frac{d\tau}{a_{\mathrm{ie}}(\tau/t_{\mathrm{ie}})^{1/2}} = 2ct_{\mathrm{ie}} \quad \text{(no inflation)}\,,$$

or about 6×10^{-26} m for our canonical example $\left(t_{\mathrm{ie}} = 10^{-34}\ \mathrm{sec}\right)$. By comparison, the horizon at the end of Inflation is

$$d_{\mathrm{PH}}(t_{\mathrm{ie}}) = c a_{\mathrm{ie}}\int_0^{t_{\mathrm{is}}} \frac{d\tau}{\tilde{a}(\tau)} + c a_{\mathrm{ie}}\int_{t_{\mathrm{is}}}^{t_{\mathrm{ie}}} \frac{d\tau}{a_{\mathrm{is}}\exp\left[H_{\mathrm{I}}(\tau - t_{\mathrm{is}})\right]} \quad \text{(inflated)}, \tag{16.4}$$

where $\tilde{a}$ is the pre-inflation expansion function. Assuming for the moment that the pre-inflation Universe is dominated by radiation or other ultra-relativistic particles, $\tilde{a}(t) = a_{\mathrm{is}}(t/t_{\mathrm{is}})^{1/2}$ and thus

$$d_{\mathrm{PH}}(t_{\mathrm{ie}}) = c\frac{a_{\mathrm{ie}}}{a_{\mathrm{is}}}\left\{2t_{\mathrm{is}} + \frac{1}{H_{\mathrm{I}}}\left[1 - \exp\left(-H_{\mathrm{I}}(t_{\mathrm{ie}} - t_{\mathrm{is}})\right)\right]\right\}, \tag{16.5}$$

$$\approx \left[2ct_{\mathrm{is}} + \frac{c}{N}(t_{\mathrm{ie}} - t_{\mathrm{is}})\right]\exp(N) \quad \text{(inflated)}, \tag{16.6}$$

for $N \gg 1$. For our $N = 70$ example, $d_{\mathrm{PH}} \approx 2600$ m at Inflation's end, about 29 orders of magnitude greater than the non-inflated expansion. This is typical of most plausible Inflation scenarios: Inflation increases the particle horizon distance over that of the non-inflating Universe by a factor of $\sim \exp(N)$ (at the time Inflation ends).

At any time $t > t_{\mathrm{ie}}$ following Inflation the particle horizon distance is

$$
\begin{aligned}
d_{\mathrm{PH}}(t) &= c a(t) \left[\int_0^{t_{\mathrm{is}}} \frac{d\tau}{\tilde{a}(\tau)} + \int_{t_{\mathrm{is}}}^{t_{\mathrm{ie}}} \frac{d\tau}{a_{\mathrm{is}} \exp\left[H_{\mathrm{I}}(\tau - t_{\mathrm{is}})\right]} + \int_{t_{\mathrm{ie}}}^{t} \frac{d\tau}{a_{\mathrm{ccm}}(\tau)} \right], \\
&= a(t) \left[\frac{1}{a_{\mathrm{ie}}} d_{\mathrm{PH}}(t_{ie}) + c \int_{t_{\mathrm{ie}}}^{t} \frac{d\tau}{a_{\mathrm{ccm}}(\tau)} \right],
\end{aligned} \tag{16.7}
$$

where a_{ccm} is the Concordance Cosmological Model expansion function of Chapter 15. For our canonical $N = 70$ inflation model the particle horizon distances at selected mileposts in the CCM expansion are:

Particle horizon distances

Milepost	with Inflation	without Inflation
End of Inflation $\left(t \sim 10^{-34}\text{ sec}\right)$	2600 m	6×10^{-26} m
Photon decoupling	4000 Mpc	0.26 Mpc
Current time	4300 Gpc	14.1 Gpc

The exact values of particle horizons in an inflating Universe depend upon the parameters presumed for Inflation, but the above values are representative of likely models. It is apparent that all of the currently visible Universe is – and always has been – within the inflated horizon.

Flatness

The Expansion Equation (9.13) may be written as $H^2(\Omega - 1) = K_0 c^2/a^2$, where $\Omega = \varepsilon/\varepsilon_{\mathrm{c}}$ includes radiation, matter, and dark energy. This can be re-arranged as

$$
\dot{a}^2(\Omega - 1) = K_0 c^2 . \tag{16.8}
$$

The quantity $\dot{a}^2(\Omega - 1)$ is thus an invariant during expansion. Written in non-parametric form for the early, radiation-dominated Universe:

$$
\frac{8\pi G}{3c^2} \frac{\varepsilon_{\mathrm{r},0}}{a^2} - \dot{a}^2 = K_0 c^2 . \tag{16.9}
$$

Both terms on the left increase without bound as $t \to 0_+$, but their difference remains constant and relatively small. From the SN Ia data there is no evidence of significant curvature on scales $\lesssim$ 4 Gpc, so the Universe's radius of curvature must be at least this large and thus $\left|K_0 c^2\right| = |c/R_0|^2 \lesssim 6 \times 10^{-3}\ \mathrm{Gyr}^{-2}$ or $\sim 10^{-38}\ \mathrm{sec}^{-2}$. Since $a(t) \sim 10^{-10}\, t^{1/2}$ for t in seconds (Equation (11.8)) and $\varepsilon_{\mathrm{r},0} \sim 10^{-13}\ \mathrm{J/m^3}$, the two terms on the left of this equation are each on the order of $10^{-20}\ \mathrm{sec}^{-2}$ at $t \approx 1$ second, and thus must balance each other to within 1 part in $\sim 10^{18}$. This is a remarkable bit of fine tuning that must be even finer at earlier times. The 'flatness' problem is thus more properly seen as one of a very fine tuning between expansion rate and energy density.

Such a fine tuning could be viewed in terms of initial conditions for the Universe's expansion but, as with the horizon problem, it is more satisfactory to find a plausible cause without invocation of special initial conditions. Inflation increases the expansion factor a, and thereby the radius of curvature R, by a factor of $\exp(N)$; which thus decreases the geometric curvature $|K| = R^{-2}$ by a factor of $\exp(2N)$. The effect is quite similar to inflating a balloon. Unless the Universe was *very* highly curved prior to inflation, it will be very flat afterward.

The resulting fine-tuning between expansion rate and energy density is again shown in Equation (16.9). From Equation (16.8) the dynamical energy parameter $\Omega - 1$ must thus change during inflation according to

$$[\Omega - 1]_{t=t_{\text{ie}}} = \frac{\dot{a}\,(t_{\text{is}})^2}{\dot{a}\,(t_{\text{ie}})^2}\,[\Omega - 1]_{t=t_{\text{is}}} \ .$$

Since $a\,(t) = a\,(t_{\text{is}}) \exp\left[H_{\text{I}}\,(t - t_{\text{is}})\right]$ during inflation, $\dot{a}\,(t) = H_{\text{I}}\,a\,(t)$ so

$$\frac{\dot{a}\,(t_{\text{is}})}{\dot{a}\,(t_{\text{ie}})} = \frac{a_{\text{is}}}{a_{\text{ie}}} = \exp(-N) \ .$$

It follows that

$$[\Omega - 1]_{t=t_{\text{ie}}} \approx \exp(-2N)\ [\Omega - 1]_{t=t_{\text{is}}} \ .$$

This typically represents a very large decrease in $|\Omega - 1|$ during inflation, leading to the current apparent fine-tuning between expansion rate and mass/energy density pretty much irrespective of the situation prior to inflation.

Monopoles

The Standard Model of particle physics predicts the creation of a large population of magnetic monopoles during the GUT phase transition when the strong nuclear force separated from the electroweak force at $t \sim 10^{-35}$ seconds; as topological defects arising from spontaneous symmetry breaking. These would be so massive that, if at all numerous, they would have long since closed the Universe and caused it to re-contract. Exactly how many monopoles were created is a matter of conjecture, but particle densities comparable to those of baryons seem plausible. Thus far, however, not a single magnetic monopole has been observed – and the Universe has certainly not collapsed.

Inflation by a factor of $\exp(N)$ dilutes particle densities by $\exp(3N)$ over what would have been the case with no inflation. The consequences for the current Universe hinge upon the densities present at the start of inflation, which is presumed to be the time of GUT symmetry breaking. In the case of topological defects attendant upon symmetry breaking, such as magnetic monopoles, an apparently reasonable minimal estimate of that primordial density is: one defect per particle horizon. If, as current quantum field theories predict, the GUT transition time was $t_{\mathrm{i}} \approx 10^{-35}$ seconds, and if the Universe up until that time was radiation-dominated, the mean distance between monopoles would have been $d_{\mathrm{PH}} = 2ct_{\mathrm{i}} \approx 10^{-26}$ meters. In our $N \approx 70$ example the GUT time would have corresponded to $a_{\mathrm{i}} \approx a_{\mathrm{ccm}}(t_i) \exp(-N) \approx 10^{-57}$, so the mean monopole separation would now be $(a_0/a_i) \times 10^{-26} \approx 10^{31}$ meters, or $\sim 10^5$ Gpc. But the current proper radius of the visible Universe is only about 14 Gpc, so the expected number of monopoles in our visible Universe is much less than 1. In effect, inflation sweeps primordial topological defects completely beyond our current horizons, thus accounting for their apparent absence.

16.1.3 Seeds of structure

The growth of gravitating structures – galaxies, clusters – in the Universe requires initial matter density perturbations to seed such growth. But in the simplest picture the very early Universe would have been homogeneous to a high degree, a consequence of very high interaction rates amongst particles and radiation in such a hot, dense state; preventing the gravitational growth of the structure currently seen. Some unusual condition must have arisen to allow density perturbations to come into existence and survive into the later Universe where they could serve as the needed seeds for structure formation.

That necessary condition appears to have been the exponential expansion associated with inflation. Quantum vacuum energy density fluctuations are a common occurrence in fields where random fluctuations occur as a matter of course; under normal circumstances such fluctuations are quickly extinguished by further random events. But if the medium is expanding at a sufficiently rapid rate, the fluctuations can expand to become much larger than their horizons and thus be immune to annihilation by normal means. If inflation is of sufficient extent, random fluctuations corresponding to cosmologically interesting sizes (i.e., that will grow to such sizes due to expansion $a(t)$) can remain well outside their horizons until after Inflation ends, and long enough for gravitation to begin their compression into the seeds of structure seen today. The growth of such density perturbations is considered further in Section 17.1.

16.1.4 Inflation realities

Speculation about possible hyper-expansion eras early in the Universe arose naturally during the 1960s and 1970s from consideration of possible scalar fields associated with the particle physics at high energies, but were not taken seriously until Guth (1981) connected the dots and proposed the Universe's homogeneity and flatness as consequences of such expansion. Even then the idea seemed a bit *ad hoc*, a theory designed to explain one thing without supporting connections elsewhere or a means of falsification (a *sine quo non* for viable theories, especially highly speculative ones). But theories of the growth of density perturbations – of as-then unknown origin – into seeds of large-scale structure in the Universe were also in the air at about the time of Guth's publication, so the realization that inflation could provide the requisite perturbations was seized upon to promote Inflation to the status of a foundation piece for modern cosmology – where it remains today.

But a coherent theory of Inflation remains elusive, largely because of our lack of a good theory for the very early Universe in which fundamental forces were joined together, and from which they separated in the TOE and GUT transitions that may have entailed inflation. The version of inflation originally proposed by Guth included no stopping mechanism, and later efforts to provide one (principally by Andrei Linde) suffered from inherent instabilities. Linde's chaotic inflation model, for instance, proposes that inflation stops locally but continues globally, so that the 'universe' becomes eternally inflating but with bubbles of 'normal space' expanding according to the Field Equations. This leads rather naturally to the idea of 'multiverses', currently enjoying some popularity among theorists: our Universe is but one of many, all disconnected from one another by the intervening inflationary expansion and (perhaps) each with its own version of fundamental physics.

For the most part, detailed theories of Inflation rely upon comparison of predicted results with those observed with the most modern instruments, particularly of the structure of the CMB radiation and of gravitational radiation (yet to be observed). At a fundamental level, Inflation remains a speculative theory, the evidence for which is not entirely conclusive in that most of it could arise from something other than inflation.

Physicists tend to accept the idea of Inflation not only because of the evidence discussed here, but also because such a mechanism relieves the theorist of the need to invoke special initial conditions to the expansion of the Universe. Evidence strongly supports a picture in which the Universe at an early stage must have been incredibly flat and very (but not completely) homogeneous. Rather than having to begin in such a special state, the Universe is essentially forced into it by Inflation, which provides the needed properties pretty much irrespective of the Universe's state at $t = 0$. It is aesthetically pleasing to carry this one step further and to propose (as many have) that the Universe *started* with Inflation; i.e., containing nothing but the inflaton field which then inflated and led to the expanding Universe that we now observe. This would have somewhat changed the timing of things in the very early Universe, and consequent horizon sizes, from those predicted by the CCM; but would not have greatly changed the later Universe in which we live.

Resolution of such issues will probably only come with development of a quantum theory of gravitation. Inflation is of great interest to physicists in part because it promises a natural conjunction of quantum physics with gravitation – the two dominant, and largely separate, paradigms underlying modern physics. One prediction of Inflation is that of gravitational waves that should have left their imprint on the CMB in the form of polarization, so astronomers are searching for these (very weak) signals as clues to the (hoped-for) quantum foundations of gravitation. As of this writing there were promising signs of such signals, but without confirmation. Stay tuned.

16.2 Post-Inflation expansion

Following Inflation the Universe is a hot plasma that expands and cools, and in the process hosts a series of particle interactions that evolve the Universe's contents. Naively extrapolating current matter densities back to the time Inflation ended suggests a matter density greater than present by a factor of $a_{\rm ie}^{-3} \sim 10^{81}$, or $\rho_{\rm m} \sim 10^{53}$ kg/m^3. This is about 37 orders of magnitude greater than nuclear densities, so we cannot expect that anything like the Universe's current contents existed at such early times. Since energies following Inflation were as high as 10^{15} GeV, ~ 13 orders of magnitude greater than can be produced in today's particle accelerators, any detailed description of the particle physics at very early times would not be credible. To the extent we can usefully speculate on such matters, the story of the evolution of the Universe following Inflation (as presented in this section) ranges from highly speculative just after Inflation ends at $t \sim 10^{-34}$ seconds, to fairly well understood following quark confinement in hadrons near $t \sim 10^{-5}$ seconds. The portion of the Universe's evolution that is dominated by elementary particles ends with the nucleosynthesis of light elements at $t \sim 1000$ seconds, which we take here to be the beginning of the Plasma Era (next chapter).

16.2.1 Baryogenesis

Early on in the post-inflation Universe the particle content probably consisted of free quarks, photons, various leptons, and – possibly – a heavy, non-relativistic particle that no longer exists. That particle – generically known as the X-particle – was endowed with mass by the spontaneous symmetry breaking of the GUT transition and probably decayed rapidly to less massive, non-relativistic particles. While it existed the X-particle and its anti-particle, $\bar{X}$, may have been responsible for the creation of a net baryon number via the process known as *Baryogenesis*.

We expect that particles and their anti-particles would have initially been created in equal numbers by decay of the inflaton field, but the current Universe appears to be populated almost entirely by particles: whatever anti-particles initially existed have been annihilated by reactions such as $x + \bar{x} \rightarrow \gamma + \gamma$, leaving a photon gas as their legacy. The particles in the current Universe thus represent a primordial excess of particles over anti-particles, that must have been created by the Baryogenesis process. The reactions involved may have been of the form[3]

$$X \rightarrow q + q\,, \quad \bar{X} \rightarrow \bar{q} + \bar{q}\,,$$

where q is a generic quark. These reactions clearly do not conserve baryon numbers so that, if the first goes on more rapidly than the second, a net excess of quarks over anti-quarks will be created, leading eventually to an excess of baryons over anti-baryons. The non-conservation of baryon numbers in such reactions is one of three conditions proposed by Andrei Sakharov in 1967, together with violation of CP (charge conjugation–parity) symmetry and departures from equilibrium, as being necessary for Baryogenesis.[4]

The resulting baryon/anti-baryon number ratio (or, more properly, the hadron/anti-hadron ratio) would have been nearly unity with a baryon excess of 1 part in $\sim 10^9$. Following quark confinement into hadrons at $t \sim 10^{-5}$ seconds, hadrons – including baryons – annihilated their anti-particles via such reactions as $x + \bar{x} \rightarrow \gamma + \gamma$, leaving $\sim 10^9$ photons per surviving baryon as a legacy of the very early Universe. That ratio of photons to baryons remains (with minor modifications) into the current era.

16.2.2 Particle equilibria and freezeouts

So long as mean particle energies remain sufficiently high, equilibration reactions of the sort

$$x + \bar{x} \rightleftarrows \gamma + \gamma \tag{16.10}$$

can take place to maintain equilibrium between a particle type (x) and its anti-particle ($\bar{x}$), where γ represents a photon (as an example). Once mean energies fall below the particle rest energy ($m_x c^2$) the pair-production half of this reaction, producing particles and anti-particles from photons, slows and eventually anti-particles are all converted to photons, leaving behind a fossil abundance of particles x without accompanying anti-particles $\bar{x}$.

More complicated reactions involving several particles, such as $a+b \rightarrow c+d$, can go forward if the reaction rates exceed the Hubble rate characterizing the

[3] Following Carrol and Ostlie (2007), p. 1246.

[4] See, e.g., Section 9.2 of Bernstein (1995) for a comprehensive and accessible discussion of the physics of Baryogenesis.

expansion; i.e., if there is time for the reaction to proceed before the Universe expands too much. Generically, reaction rates go as

$$\Gamma_{ab} = n_{a,b}\sigma_{ab}c \,,$$

where n is the particle density and σ is the reaction cross-section. The density typically varies with expansion as $n \propto a^{-3} \propto t^{-3/2}$, while reaction cross-sections are usually proportional to a high power of $\mathcal{E} \propto t^{-1/2}$; so reaction rates typically fall rapidly with time. Meanwhile, the Universe's expansion rate goes as $H \propto t^{-1}$ (Equation (11.9)) so that, as expansion goes on, the reaction rate Γ will typically fall below the Hubble rate and the reaction will effectively stop. When this happens depends critically on the behavior of the cross-section with particle energy.

An important application of these ideas involves the reactions mediating proton and neutron numbers. At sufficiently high energies, neutrons (N) and protons (P) are in equilibrium via the reactions

$$N + e^{+} \rightleftarrows P + \bar{\nu}_e \,, \tag{16.11}$$

$$P + e^{-} \rightleftarrows N + \nu_e \,, \tag{16.12}$$

where e^{+} is the positron, the anti-particle of the electron $\left(e^{-}\right)$; and ν denotes a neutrino (similar equilibrating reactions take place with electrons and their neutrinos replaced with the tauon and muon leptons and their corresponding neutrinos). So long as the reaction rates are large in comparison with the Hubble rate, the relative abundances of neutrons and protons are the thermodynamic equilibrium ones given by the Boltzmann factor,

$$\frac{n_{\rm N}}{n_{\rm P}} \approx \exp\left(-\frac{Q}{k_{\rm B}T}\right) \,, \tag{16.13}$$

where $Q = 1.29$ MeV is the excess of neutron rest energy over that of the proton. From this and Equations (11.1) and (11.8),

$$\frac{n_{\rm N}}{n_{\rm P}} \approx \exp\left(-t^{1/2}\right) \,, \tag{16.14}$$

for t in seconds.

The equilibrating reactions (16.11), (16.12) are mediated by the weak nuclear force (WNF) and thus are very sensitive to temperature. At energies just less than ~ 1 MeV the neutrino reaction rates fall below the Hubble rate in the CCM and the neutron–proton ratio freezes out at $n_{\rm N}/n_{\rm P} \sim 0.2$; the corresponding time is $t_\nu \approx 2$ seconds, and the temperature is $\sim 10^{10}$ K. Subsequently, free neutrons β-decay relatively slowly to protons by $N \rightarrow P + e^{-} + \bar{\nu}_e$ with an e-folding time scale of $\tau_{\rm N} \approx 887$ seconds, or a half-life of ≈ 615 seconds.

As another example, neutrinos (ν) are initially coupled to baryonic matter by such WNF reactions as those in Equations (16.11) and (16.12), and

$$\nu + \bar{\nu} \rightleftarrows e^- + e^+ ,$$
$$\nu + e^- \rightleftarrows \nu + e^- , \quad \text{etc.}$$

All such reactions effectively stop when the mean particle energy falls below that required for WNF reactions ($\sim$ 0.8 MeV) and neutrinos effectively decouple from matter.

Subsequent to neutrino decoupling, the mean particle energy falls below that required for $e^- - e^+$ pair production ($\sim$ 0.5 MeV at $t \approx 30$ seconds) and positrons are annihilated via $e^- + e^+ \rightarrow \gamma + \gamma$, a process that re-heats the Universe and raises photon and baryon temperatures above those of the already decoupled neutrinos. The result is a primordial neutrino population with a lower temperature and energy density than those of primordial photons. Detailed calculations[5] predict a primordial neutrino energy density 68% that of primordial photons, so that the energy density of relativistic particles is $\approx$ 1.68 times that of the CMB photons alone. It is this number – $1.68\, a_{\rm rad} T_0^4 \approx 7.01 \times 10^{-14}$ J/m^3 – that is normally taken as the current CMB energy density.

Immediately following positron annihilation the Universe's (nearly) stable contents are: protons and neutrons, electrons, neutrinos, and photons. These form an ionized gas (plasma) whose mean particle energy is on the order of several hundred keV and whose matter density is $\rho_{\rm m,0}/a^3 \sim 1$ kg/m^3, roughly that of air under standard (terrestrial!) conditions. The stage is now set for this expanding gas to form the structures we see in the current Universe. That process begins with assembly of protons and neutrons into light atomic nuclei, much as happens in the cores of main sequence stars; but this primordial nucleosynthesis takes place under conditions quite different from those in stars.

16.3 Primordial nucleosynthesis

The foundation reaction for primordial nucleosynthesis is that of deuterium formation:

$$P + N \longrightarrow {}^2\text{H} + \gamma \, . \tag{16.15}$$

Cross-sections for this reaction are so large that even at modest particle energies it would go forward very rapidly.[6] But deuterium's nucleus is a fragile one, with

[5] See, e.g., Section 9.6.5 of Ellis *et al.* (2012), Section 5.3 of Narlikar (1993), or Section 4.2.1.7 of Peter and Uzan (2009).

[6] In contrast, the foundation reaction in solar-mass stars, $P + P \rightarrow {}^2\text{H} + e^- + \bar{\nu}_{\rm e}$, goes forward much too slowly to contribute to primordial nucleosynthesis.

a binding energy of ~ 2 MeV; and is easily photodissociated by sufficiently energetic photons. The number density of photons exceeds that of baryons by a factor of $\sim 10^9$ (Section 11.1) so that, even at a temperature less than that corresponding to the deuterium binding energy, plenty of photons would be available to dissociate any newly formed deuterium nucleus and we would expect the equilibrium deuterium abundance to be small. For primordial nucleosynthesis to proceed we must wait for the Universe to cool to the point that deuterium can survive; or, more precisely, for the nucleosynthesis rate to exceed that of photodissociation.

But we cannot wait too long, for free neutrons are β-decaying with a half-life of $\sim$ 10 minutes and, as time goes on and the Universe cools, the Coulomb barriers of light nuclei become harder to overcome. As a result there is a restricted window in the Universe's expansion during which nucleosynthesis may occur, and an upper limit to the atomic mass of elements formed then. For the CCM expansion history and the Standard Model of particle physics, the nucleosynthesis window is nominally between ~ 100 and ~ 1000 seconds and the heaviest nucleus formed in any abundance is helium-4 (^{4}He).[7] By the time ^{4}He has formed the mean particle energy in the Universe is too small for heavier nuclei – entailing larger Coulomb barriers – to form in any quantity. This is quite unlike the case with stars where mean energies are *increasing* as time goes on, leading to progressively heavier nuclei.

By the time nucleosynthesis of deuterium has begun in earnest, the N/P ratio has dropped to $\sim$ 1/7 so there are two neutrons for every 16 baryons. Nearly all of those neutrons will wind up in ^{4}He nuclei since that is the most tightly bound of all the light compound nuclei; and thus there will be about one helium nucleus (four baryons) for every 16 baryons total, or a ^{4}He mass fraction of $\sim$ 0.25. And, indeed, this is approximately what is observed in today's Universe at large.

But looked at more closely, the exact value of this mass fraction is dependent upon the starting time for deuterium nucleosynthesis, which in turn is dependent upon the photodissociation rates in the early Universe. These rates are nearly proportional to the number density of energetic photons; or, more precisely, to the photon/baryon number ratio, η^{-1}. Observations of the abundances of primordially formed light elements thus can reveal this number ratio which, in combination with the known photon number density of the dominant CMB, will tell us the

[7] The standard notation for isotope nuclei is ^A_ZX for the isotope of element X containing A nucleons (protons plus neutrons). The subscript Z denotes the number of protons so that the neutron number is $A - Z$. The element name X is a classical proxy for Z so that this atomic number is redundant and often omitted. Thus, ^{4}He = $^4_2$He is the isotope of the two-proton element (He) containing $4 - 2 = 2$ neutrons, $^{235}_{92}$U is the isotope of uranium ($\Rightarrow$ 92 protons) with 143 neutrons, etc. The two heavy isotopes of hydrogen are traditionally given their own names: deuterium for ^{2}H and tritium for ^{3}H.

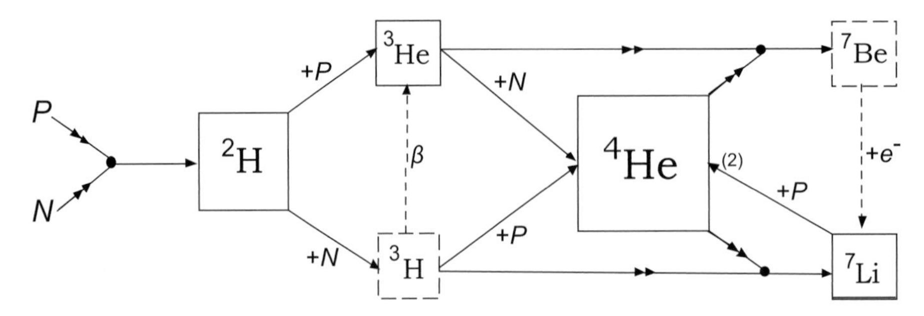

Figure 16.2 Principal reaction pathways in primordial nucleosynthesis. Dashed boxes denote unstable isotopes that later decay by the dashed pathways shown (see text).

baryon number density. This, as it turns out, is the most convincing evidence for our current estimate of the baryonic contents of the Universe.

The principal routes to primordial nucleosynthesis of light elements are shown in Figure 16.2. These reactions begin in earnest at $t \sim 100$ seconds when there is about one deuterium nucleus (^{2}H) per million ordinary hydrogen nuclei (P), and continue until the supply of neutrons is exhausted and temperatures have fallen too far for charged particles to combine. Elements heavier than helium cannot form in any quantity due to the absence of stable isotopes of atomic numbers 5 and 8 and because falling temperatures render the larger Coulomb barriers largely impassible; but traces of ^{7}Li and ^{7}Be are formed at levels several orders of magnitude less than that of deuterium, and *very* small amounts of ^{6}Li, ^{9}Be, and isotopes of boron and carbon are also formed. By about $t \sim 1000$ seconds the neutron supply is nearly exhausted and nucleosynthesis has largely ceased; a cosmologically short time later, the unstable isotopes ^{7}Be and ^{3}H (tritium) decay by the pathways shown (as dotted lines) to the stable isotopes ^{7}Li and ^{3}He, respectively. The eventual result leaves about one quarter of the Universe's baryonic mass in the form of ^{4}He and $\sim 10^{-5} - 10^{-4}$ each in the form of ^{2}H and ^{3}He, and a substantially lesser amount of ^{7}Li.

The progress of these reactions can readily be modelled by sets of linked, first-order differential rate equations. The evolution of deuterium (D) abundance, for instance, is described by

$$\frac{d}{dt}n_{\mathrm{D}} = A \cdot n_{\mathrm{P}} \cdot n_{\mathrm{N}} - B \cdot n_{\mathrm{P}} \cdot n_{\mathrm{D}} - C \cdot n_{\mathrm{N}} \cdot n_{\mathrm{D}} - E \cdot n_{\mathrm{D}} \,, \tag{16.16}$$

together with similar – and linked – rate equations for P, N, ^{3}H, and ^{3}He; where n denotes number densities and the rate coefficients (A, B, C, E) are functions of T (hence of t), known from particle accelerator experiments.[8] The second and

[8] More precise computations employ second-order reactions – described by additional terms – in addition to those in Equation (16.16); and similarly for the evolution of other nuclear abundances.

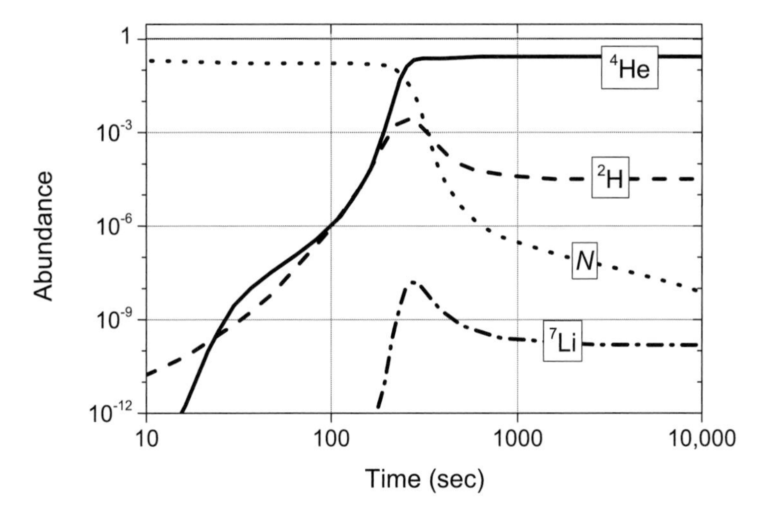

Figure 16.3 Abundance (mass fraction) evolution of key isotopes during primordial nucleosynthesis (N = neutrons).

third terms on the right (B, C) denote deuterium destruction in the formation of ^{3}He and ^{3}H, respectively; and the last term (E) models photo-dissociation of deuterium. Simultaneous numerical solutions of the complete set of differential equations for all isotopes, whose rate coefficients embody the CCM and presumed baryon/photon ratio, yield theoretical timelines for light nuclei productions and their final abundances as partially illustrated in Figure 16.3. Decreases in ^{2}H and ^{7}Li after $\sim$ 250 seconds reflect their consumption in the production of ^{4}He.

16.3.1 Nucleosynthetic yields

The light element abundances from primordial nucleosynthesis are sensitive to the value of η or, if you like, to the number of nucleons available and of photodissociating photons per participating nucleon. In particular, the neutron and proton number densities are directly proportional to η for given photon number densities, and photodissociation rates are inversely proportional to η; so the deuterium yield is a contest between production and destruction, mediated by the baryon–photon ratio. The same is true (more or less) for the other light nuclei. As a consequence, ^{4}He yields are larger for larger values of η; deuterium and ^{3}He yields *decrease* with increasing η as these isotopes are increasingly converted to ^{4}He; and ^{7}Li yields vary with η in a non-monotonic manner (due to its complex nucleosynthesis pathways). For the canonical $H_0 \approx 0.074$ Gyr^{-1} and the Standard Model of particle physics (which entails three lepton families), the primordially produced abundances of the four most useful light isotopes (those enclosed in solid lines in Figure 16.2) are given by

$$\frac{n\{^4\mathrm{He}\}}{n\{^1\mathrm{H}\}} \approx 0.165 + 0.009 \log \eta \ , \tag{16.17}$$

$$\frac{n\{^2\mathrm{H}\}}{n\{^1\mathrm{H}\}} \approx 5.5 \times 10^{-20} \eta^{-1.6} \ , \tag{16.18}$$

$$\frac{n\{^3\mathrm{He}\}}{n\{^1\mathrm{H}\}} \approx 1.8 \times 10^{-11} \eta^{-0.63} \ , \tag{16.19}$$

$$\frac{n\{^7\mathrm{Li}\}}{n\{^1\mathrm{H}\}} \approx 1.1 \times 10^{-33} \eta^{-2.38} + 2.8 \times 10^{12} \eta^{2.38} \ . \tag{16.20}$$

The complicated result for ^{7}Li reflects the relative complexity of its synthesis, with two formation channels (via ^{3}H and ^{7}Be) and one destruction channel (to ^{4}He). See, e.g., Section 4.3.4 of Peter & Uzan (2009) for details of primordial nucleosynthetic reactions for all the light elements.

Comparison of the most important of these theoretical results with observations is shown in Figure 16.4, where the dotted horizontal lines delimit observational estimates of abundances. Abundance estimates are complicated by the

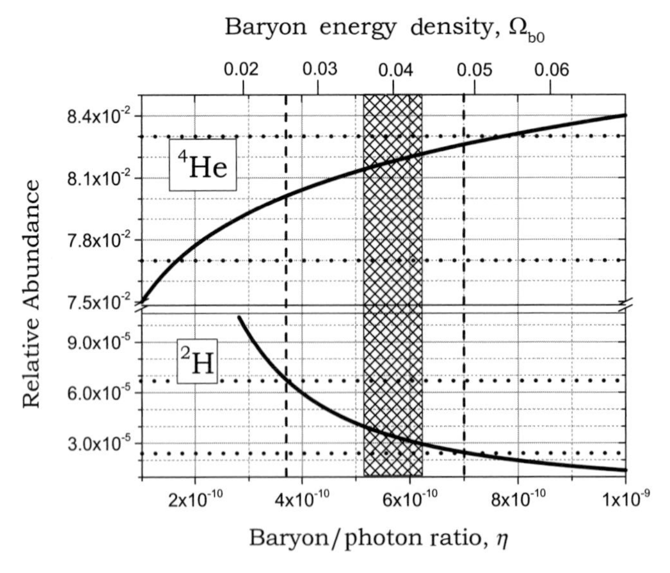

Figure 16.4 Primordial nucleosynthetic yields (number relative to ^{1}H) of deuterium (^{2}H) and helium (^{4}He) as functions of the baryon/photon number ratio or, equivalently, the baryon energy density (see text). Horizontal dotted lines show the range of observed abundances; vertical dashed lines show the corresponding range of baryon/photon number ratio. Note abundance axis break between the two isotopes.

usual uncertainties in spectroscopic abundance analyses, exacerbated by the need to restrict the resulting abundance estimates to primordial ones.[9] All the light compound nuclei other than that of ^{4}He are weakly bound and easily destroyed in stellar interiors, while ^{4}He is manufactured in stars; so observed abundances are often other than primordial and do not directly reflect primordial nucleosynthesis. These complications result in fairly wide bounds for primordial abundance estimates, as illustrated by the horizontal dotted lines in Figure 16.4.

The estimated limiting values for η, consistent with observations of both ^{2}H and ^{4}He – the most useful isotopes for this purpose – lie in the approximate range 4×10^{-10} to 7×10^{-10} (dashed vertical lines); the conventional best estimates for η lie between $\sim$ 5 and $\sim$ 6 times 10^{-10} (hatched regions). The resulting range of values for the current baryon mass/energy density parameter, Ω_{b0}, are between ~ 0.034 and ~ 0.042 . Similar analyses of ^{3}He give roughly concordant results but with relatively large uncertainties in primordial corrections. Observed abundances of ^{7}Li are consistently less than those predicted from analyses of other light elements by factors of 2 to 4; the reason for this discrepancy is unknown. Irrespective of this, the current baryon energy density appears to lie in the range $0.03 \lesssim \Omega_{b,0} \lesssim 0.05$, a robust conclusion with profound consequences.

16.3.2 Baryonic mass density estimates

The canonical estimates for baryonic matter from these analyses are $\eta \approx 5.5 \times 10^{-10}$ and $\Omega_{b,0} \approx 0.04$. This last number is both too large and too small for comfort: $\Omega_{b,0}$ is far less than the total matter density estimate from gravitational observations and model analysis $(\Omega_{m,0} \approx 0.27)$, and much larger than that of luminous baryonic matter $(\Omega_{lb,0} \approx 0.006)$. It appears that most of the Universe's matter is not baryonic in nature, and most of the Universe's baryonic matter is hiding out of sight. There is probably no way around such conclusions, based as they are on well-understood physics and a wide range of (mostly concordant) observations.

Problems

1. The Sun has been powered for its lifetime – about 4.5 billion years – by fusion of hydrogen to helium, yielding $\sim$ 27 MeV per helium nucleus produced. Assuming that the Sun's luminosity has been its current value

[9] See Boesgaard & Steigman (1985), *Ann. Rev. Astron. Astrophys.* **23**, pp. 319–378, for an extensive discussion of observational techniques for primordial light element abundances. Mo *et al.* (2010), Section 3.4.4, and Peter & Uzan (2009), Section 4.3.5, summarize current observational results.

during its lifetime (a substantial overestimate), compute how much helium has been produced inside the Sun during its lifetime, and compare that to the Sun's current helium content, assuming it – like nearly all stars – is $\sim$ 25% helium by weight. It was this sort of consideration (extended to stars as a whole) that motivated the original proposal for a hot beginning for the Universe, well before there was any specific evidence for such a proposal (such as the CMB).

2. Primordial nucleosynthesis injected $\sim$ 27 MeV per helium nucleus into the universal plasma. Compute the extent to which the plasma's temperature was thereby increased and comment on the importance of primordial nucleosynthesis to the Universe's mean particle energy.
3. Derive a formula relating the relative numbers of helium-4 nuclei, $n\left\{{}^{4}\mathrm{He}\right\}/n\left\{{}^{1}\mathrm{H}\right\}$, to the helium mass fraction $Y \equiv m\left({}^{4}\mathrm{He}\right)/\left[m\left({}^{4}\mathrm{He}\right)+m\left({}^{1}\mathrm{H}\right)\right]$. Verify that the He/H number ratio shown as the relative abundance in Figure 16.4 is consistent with the canonical helium mass fraction of $\sim 1/4$.
4. Show that the event horizon in an exponentially expanding Universe is given by $d_{\mathrm{EH}} = c/H$, where H is the invariant Hubble Parameter characterizing the expansion.
5. Estimate (roughly) the abundance of helium in the Universe (as a mass fraction) if (a) $Q = m\{N\} - m\{P\} \rightarrow 0.129$ MeV; and (b) the neutron β-decay half-life $\rightarrow$ 100 seconds. Comment on the utility of the observed helium abundance in the confirmation of the Big Bang theory.

17

Plasma Era

The Plasma Era nominally begins at $t \sim 1000$ seconds, when nucleosynthesis ceases; immediately thereafter the Universe consists of a hot, expanding, ionized plasma containing, for each helium-4 nucleus: ~ 12 hydrogen nuclei, ~ 14 electrons, a few times 10^9 photons and neutrinos, $\lesssim 10^{-4}$ nuclei of various light isotopes; and an unknown number of WIMPs whose total mass is ~ 6 times that of baryonic matter. The era ends at $\sim 375,000$ years when electrons are captured onto positive ions ('recombination') and the gas comprising the Universe becomes electrically neutral ('photon decoupling').

The Universe during the Plasma Era is surprisingly thin, with a total matter density varying from ~ 0.01 kg/m^3 at the beginning to $\sim 10^{-18}$ at the end.[1] Mileposts for the Plasma Era are:

- at $t \sim 1000$ seconds, $\langle \mathcal{E} \rangle \sim 0.1$ MeV and nucleosynthesis has essentially ended; this marks the nominal start of the Plasma Era;
- at $t \approx 52,000$ years, $\langle \mathcal{E} \rangle \sim 2.5$ eV and matter takes over from radiation as the dominant source of gravitation: the Universe's expansion rate increases from $H \rightarrow (1/2)\,t^{-1}$ to $H \rightarrow (2/3)\,t^{-1}$;
- between $\sim 30,000$ and $\sim 100,000$ years, helium ions capture two electrons each as mean particle energies fall below helium ionization potentials; helium thereafter exists as neutral atoms;
- at $t \approx 375,000$ years, $\langle \mathcal{E} \rangle \sim 0.7$ eV ($T \approx 3000$ K) and the remaining free electrons are captured onto hydrogen nuclei, an event variously referred to as 'photon decoupling' or 'recombination'. Essentially all nucleons are then bound into electrically neutral atoms and very few electrons remain free. At this point the Universe becomes transparent and photons are thereafter

[1] By comparison, the Earth's sea level atmospheric density is ~ 1 kg/m^3, and the Sun's chromospheric density is $\sim 10^{-4}$ kg/m^3.

free to stream unimpeded as the Cosmological Microwave Background (CMB) radiation. This marks the end of the Plasma Era.

At first glance, little of consequence seems to happen in the dilute plasma as the Universe expands and cools, until its end when the plasma becomes a gas of neutral atoms. But, viewed more closely, the Plasma Era is the incubator for growth of the matter density perturbations created in the Inflationary epoch (Section 16.1.3), and leads eventually to the formation of galaxies and large-scale structure in the coming Galaxy Era.

17.1 Matter density perturbations

Following recombination, the matter density perturbations formed during Inflation, and evolved in the Particle and Plasma Eras, act as seeds for gravitational collapse of structures that will form galaxies and clusters of galaxies. To appreciate how these matter density perturbations evolve to become seeds of growth, it is best to begin at the end of the story, in the times just preceding recombination. At such times the perturbations are entirely contained within the particle horizon and are thus able to dynamically evolve as coherent structures. Cosmologically relevant matter perturbations (proper radii $\lesssim$ 150 Mpc) are also much smaller than the Universe's radius of curvature $\left(R_0 \gtrsim 4\text{ Gpc}\right)$ so their dynamics can be investigated with Newtonian physics. Those dynamics are complicated by the presence of two distinct types of matter, baryonic and dark. The former is closely tied to radiation by photon scattering off electrons, so the evolution of structures in this three-component fluid is complicated, but readily modelled with known physics.

But, earlier in the Universe's expansion, matter perturbations of relevant sizes were larger than the horizons at the time so that their evolution cannot be studied with Friedmann models, greatly complicating the analysis. Even earlier – during Inflation itself – the statistical properties of the perturbation population were established by *a priori* unknowable means so that we cannot independently establish what mixture of perturbation masses were created then.

Matter perturbations are usefully characterized by their linear size (length, diameter) and perturbation magnitude. The latter is commonly expressed in relative terms:

$$\delta_{\rm m} \equiv \frac{\rho_{\rm m} - \bar{\rho}_{\rm m}}{\bar{\rho}_{\rm m}} , \tag{17.1}$$

where $\bar{\rho}_{\rm m}$ is the mean (i.e., spatially averaged) density. It follows that

$$\rho_{\rm m} = \bar{\rho}_{\rm m} \left(1 + \delta_{\rm m}\right) . \tag{17.2}$$

The density perturbations created during Inflation evolve into the CMB radiation anisotropies observable in microwaves as temperature differences, as seen in Figure 17.4. The theoretical cosmologist's burden is to understand how primordial perturbations emergent from Inflation can evolve into these anisotropies, and so deduce important properties of the Universe – such as curvature and dark matter content. This is one of the most difficult and rewarding challenges in modern cosmology, which we can only sketch in this chapter.

17.1.1 Early evolution

Primordial characteristics

We do not know the physics of Inflation well enough to confidently predict the statistical properties of density perturbations created therein, so we must speculate on reasonable bases and test the consequences against observations. Perturbation sizes (L) are almost certainly broadly distributed since they reflect inflationary extents over the period of inflation; to first order, all values of L are approximately equally represented amongst primordial perturbations. Perturbation density magnitudes (δ_{m}), on the other hand, are poorly constrained by known physics. Based on scaling arguments, Harrison and Zel'dovich propose a power-law distribution[2] for perturbation density magnitudes, which are effectively scale-free in that they imply no preferred density perturbation scale. It is conventional to express the density amplitude in terms of Fourier expansions:

$$\frac{\delta\rho\,(\mathbf{x})}{\rho} = (2\pi)^{-3} \int \delta_k \exp\left(i\mathbf{k}\cdot\mathbf{x}\right)\, d^3k \,. \tag{17.3}$$

In the Harrison–Zel'dovich scaling, the coefficients scale as $|\delta_k|^2 \propto k^n$ and[3]

$$\frac{\delta\rho}{\rho} \propto M^{-n/6-1/2} \,, \tag{17.4}$$

where M is the mass of the perturbation. The coefficient n is the spectral index and takes on different values in different contexts.

Super-horizon evolution

For the purpose of following matter perturbation evolution immediately following inflation, it is best to use the Hubble horizon as the relevant causal barrier and to start the clock running at the end of inflation, when the particle horizon distance

[2] Harrison, R. (1970), *Phys. Rev. D* **1**, p. 2726; Zel'dovich, Ya. B. (1972), *Mon. Not. Royal Astron. Soc.* **160**, p. 1P.

[3] Kolb and Turner (1990), Section 8.1

is effectively zero. Matter perturbation sizes will grow as $L(t) = L_0 a(t)$, where L_0 is the co-moving length; i.e., that at the current time (assuming no internal changes). Adopting for purposes of illustration the small-cluster diameter 1 Mpc as L_0 and confining ourselves to the radiation era where $a(t) \sim 2 \times 10^{-10} t^{1/2}$ (Equation (11.8)), the growth of the perturbation's diameter following Inflation would be given approximately by $L(t) \approx 6 \times 10^{12} t^{1/2}$ meters (for t in seconds, and assuming our canonical inflation model with $a_{\mathrm{ie}} = 10^{-27}$). The Hubble horizon, on the other hand, will grow approximately as $d_{\mathrm{PH}}(t) \approx ct \approx 3 \times 10^8 t$ (for $t \gg t_{\mathrm{ie}}$ in seconds); so that at, say, $t = 1$ second the perturbation diameter will be ~ 4 orders of magnitude greater than the horizon size.

The parlance for such a situation is 'super-Hubble', or (we prefer) 'super-horizon'. Essentially, this sample perturbation ($L_0 = 1$ Mpc) will contain $\sim 10^{12}$ horizon-limited regions within it at the age of 1 second, each of them evolving independently of the others. But the horizon is growing faster than the perturbation so after a time given by $6 \times 10^{12} t^{1/2} = ct \Rightarrow t \approx 10^8$ seconds, or ~ 3 years, the horizon will have grown to encompass the entire perturbation, which can subsequently evolve coherently. The parlance here is 'the perturbation enters the horizon'; what is actually happening is that the horizon is enveloping the perturbation. But the result is the same.

Dynamics in super-horizon situations can be terribly complicated and difficult, involving solutions of perturbed Einstein Equations in circumstances where parts of the structure do not share the same LIRF as others. This is, arguably, the most complex and difficult analysis problem in cosmology, and one that we can only sketch here. The ambitious student may want to consult Liddle and Lyth (2000), Kolb and Turner (1990), Longair (2008), Peter and Uzan (2009), among others, for the details, which are non-trivial (to say the least). In what follows we summarize the conclusions from such analyses, to the extent that they appear to be concordant amongst themselves and with current observations.

17.1.2 Perturbation growth

Most density perturbations of cosmologically interesting size – i.e., corresponding to proper sizes comparable to those seen in the current Universe – remain super-horizon well into the post-inflation stages of expansion. Whatever the details of their internal evolution prior to coming into the horizon, a consequence of the Harrison–Zel'dovich scaling of their perturbation magnitudes is that *all matter perturbations have the same density magnitude δ_{m} at the time of horizon entry.* The Harrison–Zel'dovich spectral index is presumed to be $n = -3$ (in the scaling used here, and adopted from Kolb and Turner (1990)) so that, from Equation (17.4),

$$\frac{\delta \rho}{\rho} = \text{constant},$$

when the perturbation enters the horizon, irrespective of its linear size. Of course, different size perturbations will enter the horizon at different times as the horizon grows, and thus be subject to gravitational evolution for different lengths of time; so the overall perturbation population will show a range of density magnitudes δ_{m}. But this consequence of the scale-free spectrum of density magnitudes imposes a correlation between perturbation sizes and density magnitudes that will be very useful in matters to come.

Once inside the horizon, the physics governing the evolution of density perturbations is that of fluid dynamics in co-moving coordinates, from which the bulk motion is removed.[4] The relevant equations are:

$$\text{continuity:} \quad \frac{\partial \delta_{\mathrm{m}}}{\partial t} + \frac{1}{a}\vec{\nabla} \cdot \left[(1+\delta_{\mathrm{m}})\,\vec{u}\right] = 0\,, \tag{17.5}$$

$$\text{Euler:} \quad \frac{\partial \vec{u}}{\partial t} + H\,\vec{u} + \frac{1}{a}\left(\vec{u} \cdot \vec{\nabla}\right)\vec{u} = -\frac{1}{a}\vec{\nabla}\Phi - \frac{1}{a\bar{\rho}_{\mathrm{m}}\,(1+\delta_{\mathrm{m}})}\vec{\nabla}P\,, \tag{17.6}$$

$$\text{gravitation:} \quad \nabla^2\Phi = 4\pi G\bar{\rho}_{\mathrm{m}}\delta_{\mathrm{m}}a^2\,. \tag{17.7}$$

The first of these expresses conservation of mass, the second is the fluid dynamics equivalent of Newton's second law, and the third is the non-relativistic version of gravitation appropriate to the small scales involved. The velocity $\vec{u}$ is the object's peculiar velocity in addition to pure expansion, Φ is the Newtonian gravitational potential, and P is the pressure. These density perturbation equations are quite complex and without simple solutions, but some insight into their nature can be gained by looking at the initial growth when both $|\delta_{\mathrm{m}}|$ and $|\vec{u}|$ are small. Substituting the above representation for ρ_{m} (Equation (17.1)) into the hydrodynamic/gravitation equations and discarding all non-linear terms in δ and $|\vec{u}|$ and their derivatives yields the dispersion relation[5]

$$\ddot{\delta}_{\mathrm{m}} + 2H\,\dot{\delta}_{\mathrm{m}} = \omega^2\delta_{\mathrm{m}}\,, \qquad \text{where} \tag{17.8}$$

$$\omega^2 = \left[4\pi G\bar{\rho}_{\mathrm{m}} - (2\pi c_{\mathrm{s}}/L)^2\right]\,. \tag{17.9}$$

Here, $H = \dot{a}/a$ is the Hubble parameter, c_{s} is the speed of sound in the medium, and L is the linear size (e.g., diameter) of the perturbation. These equations will be suitable approximations whenever $|\delta| \ll 1$ and $|u| \approx 0$, as we expect to be the case early in the Universe's expansion.

Equations (17.8) and (17.9) superficially resemble those for a classical driven, damped oscillator of frequency $|\omega|$, where $2H\,\dot{\delta}_{\mathrm{m}}$ ('Hubble drag') is the damping term, $4\pi G\bar{\rho}_{\mathrm{m}}\delta$ is the gravity driving term, and c_{s}/L corresponds to a restoring pressure force. But in our context all the coefficients – H, $\bar{\rho}_{\mathrm{m}}$, c_{s}, L – are functions

[4] See, e.g., Mo *et al.* (2010), Section 4.1, for derivation of density perturbation dynamics.
[5] Mo *et al.* (2010), Equation 4.29.

of time, so the solutions are more varied and complex than is usually the case in classical mechanics.

Some insight into the nature of the resulting evolution of the perturbation over-density comes from consideration of the static universe case, where $H = 0$ and all the coefficients in Equation (17.9) are constant. Then the perturbation equation is $\ddot{\delta}_{\rm m} = \omega^2 \delta_{\rm m}$ with constant ω, the solutions to which are $\delta_{\rm m}(t) = C_1 \exp(\omega t) + C_2 \exp(-\omega t)$ for some constants C. If $\omega^2 > 0$, ω will be real and the perturbation density will (eventually) experience exponential growth; but if $\omega^2 < 0$, ω will be imaginary and the density will oscillate in amplitude. Put the damping term containing H back into the equation and you have, to first order, either damped exponential growth $(\omega^2 > 0)$ or damped oscillations $(\omega^2 < 0)$. The gross evolutionary behavior of density perturbations will thus be set by the sign of ω^2 or, in basic terms, by the magnitude of the pressure term $(c_{\rm s}/L)^2$ in comparison with that of gravitation, $G\bar{\rho}_{\rm m}$.

It's not hard to appreciate how this situation could arise. The time taken for free-fall gravitational collapse of a pressure-free gas cloud is the dynamical time, $t_{\rm dyn} \sim \sqrt{\pi/G\rho_0}$; and the time taken for a pressure (i.e., *sound*) wave to travel through the cloud is $t_{\rm sound} \approx L/c_{\rm s}$. Then from Equation (17.9),

$$\omega^2 = 4\pi^2 \left(\frac{1}{t_{\rm dyn}^2} - \frac{1}{t_{\rm sound}^2} \right)$$

and ω^2 will be positive if, and only if, the dynamical collapse time is less than the sound crossing time; i.e., if the cloud can collapse before a pressure wave can provide contravening support. To formalize this in terms familiar to students of star formation, the **Jeans length** $L_{\rm J}$ and **Jeans mass** $M_{\rm J}$ are defined as[6]

$$L_{\rm J} \approx c_{\rm s} t_{\rm dyn} \approx \left(\frac{\pi}{G\rho} \right)^{1/2} c_{\rm s} \,, \qquad (17.10)$$

$$M_{\rm J} \equiv \frac{4}{3}\pi\rho \left(\frac{L_{\rm J}}{2} \right)^3 \approx \left(\frac{\pi^5}{G^3\rho} \right)^{1/2} \frac{c_{\rm s}^3}{6} \,. \qquad (17.11)$$

Then a self-gravitating gas cloud of initial density ρ and sound speed $c_{\rm s}$ can collapse if $L > L_{\rm J}$ or, equivalently, if $M > M_{\rm J}$; otherwise it will oscillate.

Perturbation growth

The matter comprising density perturbations is (in general) of two quite different forms: *baryonic* and *dark*. During the plasma era baryonic matter is ionized and

[6] Sparke and Gallagher (2007), Section 2.4.2; named for the early twentieth-century British astrophysicist James Jeans who first introduced this form of stability analysis to the study of star formation.

strongly coupled to radiation by photon scattering off electrons, which provides a substantial stabilizing pressure against perturbation growth; while dark matter is independent of radiation and subject to no such radiation pressure, but *is* gravitationally coupled to baryons. The physics of structure growth in these two plasma components can be clarified by re-casting the perturbation equation (17.8), (17.9) in terms of the expansion function. From the parameterization introduced in Chapter 9, and defining the co-moving perturbation diameter L_0 by $L = L_0 a$, the perturbation equation is

$$\ddot{\delta} + \overbrace{2H\dot{\delta}}^{\text{Damping}} = \underbrace{\overbrace{\frac{3H_0^2\Omega_{\mathrm{m},0}}{2}\frac{\delta}{a^3}}^{\text{Driving}} - \overbrace{\frac{4\pi^2 c_\mathrm{s}^2}{L_0^2}\frac{\delta}{a^2}}^{\text{Restoring}}}_{\omega^2\delta} . \tag{17.12}$$

The nature of the growth of the density perturbation reflects the interplay of the driving and restoring terms on the right-hand side of this equation, which differs greatly between baryonic and dark matter and evolves as a increases. In particular, the speed of sound in the coupled baryon/photon fluid is so large that the restoring term typically dominates and induces oscillations in the perturbation amplitudes, while just the opposite is the case with dark matter which is largely free to grow. Since each type of matter is gravitationally coupled to the other, the resulting perturbation growth can be rather complicated.

Baryons and photons The growth of perturbations in the coupled baryon/photon fluid is greatly impeded by radiation pressure. Photons are much more numerous than baryons so the speed of sound in the fluid is dominated by radiation, for which $c_\mathrm{s} = c\sqrt{dP/d\rho} = c/\sqrt{3}$; the presence of matter slightly decreases this speed:[7]

$$c_\mathrm{s} = c\left[\frac{9}{4}\frac{\Omega_{\mathrm{b},0}}{\Omega_{\mathrm{r},0}}a + 3\right]^{-1/2} , \tag{17.13}$$

so that c_s varies from $\sim 0.58c$ to $\sim 0.50c$ as the plasma era progresses. This large speed of sound causes the restoring term in the perturbation equation to dominate the gravitational term for cosmologically relevant perturbation sizes, and baryon density perturbations thus fluctuate at frequency

$$|\omega| = \frac{2\pi c_\mathrm{s}}{L_0}\frac{1}{a} \approx \frac{\pi c}{L_0 a} \quad \text{(baryons)} , \tag{17.14}$$

[7] Mo *et al.* (2010), Equation 4.34.

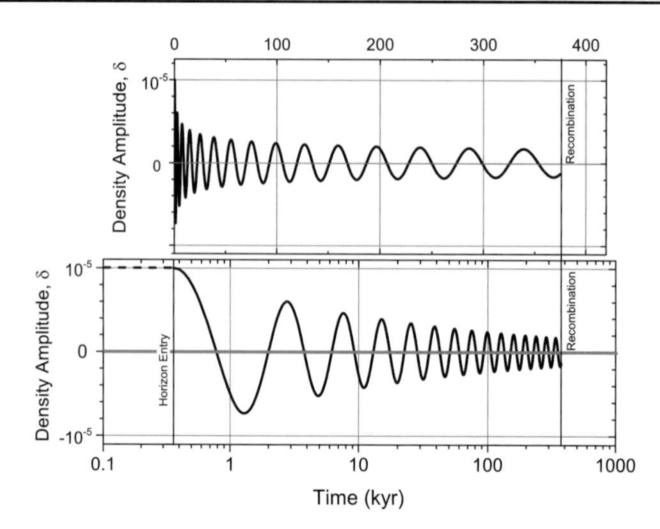

Figure 17.1 Schematic behavior of acoustic oscillations of an $L_0 = 10$ Mpc baryonic density perturbation on linear (top) and logarithmic (bottom) time scalings. The perturbation enters the horizon at $t \approx 360$ years and then begins oscillating with decreasing frequency and amplitude until recombination at 375 kyr.

after coming into the horizon, rather than continue to grow in density contrast. Thus, a typical $L_0 = 10$ Mpc perturbation (Figure 17.1) would have an oscillation frequency at the time of matter–radiation equality ($t \approx 52{,}000$ years, $a \approx 3 \times 10^{-4}$) of $\sim 10^{-11}$ s^{-1}, or an oscillation period of ~ 20 kyr. Hubble drag dissipates the amplitude and spatial expansion lessens oscillation frequency as the oscillations continue (see Figure 17.1). Note that the amplitude at photon decoupling (recombination) time is sensitive to the phasing of the oscillations, hence to the horizon entry time, hence to the linear size L_0 of the perturbation; in a non-monotonic manner.

Limitations on baryonic perturbation growth in the plasma era are illustrated in Figure 17.2. Because the baryonic Jeans length (Equation (17.10)) remains larger than the particle horizon (heavy solid line) throughout the plasma era – a consequence of the large sound speed in photon gases – a baryonic perturbation cannot simultaneously be larger than the Jeans length and be within the horizon, and so cannot grow appreciably during the plasma era. The perturbation length represented by the upper thin line in the figure – corresponding to $L\left(t_{\mathrm{pd}}\right) = 140$ kpc or $L_0 \approx 150$ Mpc, the largest structure in the current Universe – only enters the horizon at ~ 80 kyr, when the baryonic Jeans length has grown too large for the baryonic perturbation to do anything other than oscillate in density contrast. Note that a similar fate awaits any baryonic perturbation of astronomically interesting size or mass; only the time of beginning of oscillations – roughly, the time of horizon crossing (solid dots in the figure) – differs with perturbation size, with smaller perturbations entering earlier.

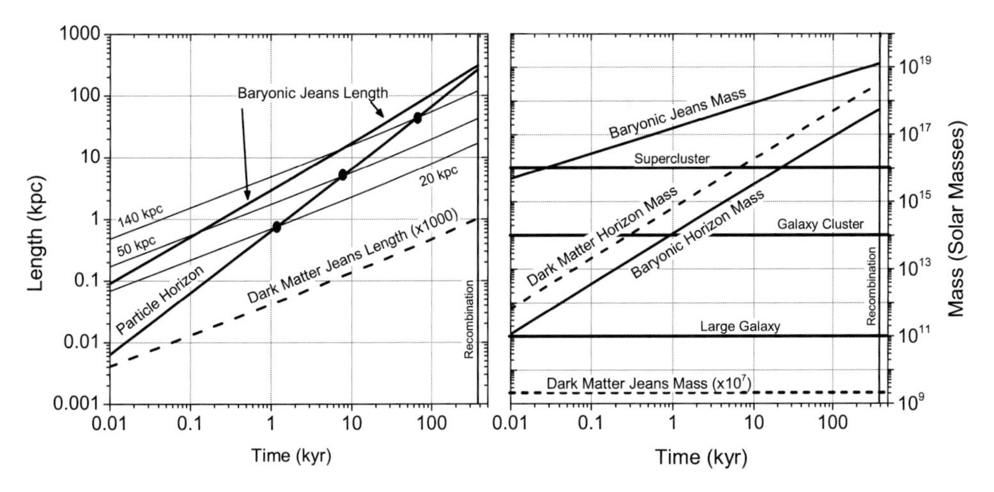

Figure 17.2 Limits to perturbation growth in the Plasma Era. The thin lines denote diameters of perturbations of labelled lengths at photon decoupling time ($\approx L_0/1090$).

Prior to photon decoupling (recombination), baryonic structures of cosmologically relevant size and/or mass were less than the Jeans length/mass and so could not coherently contract and increase their density contrast. Since current baryonic structures have $\delta_{\rm m} \gg 1$, they could not have arisen solely from baryonic perturbations in the plasma era.

Dark matter Dark matter is currently believed to be a form of matter that interacts very weakly with electromagnetism, if at all: it does not absorb, radiate, or scatter photons. It is thus impervious to radiation pressure, so the pressure represented by the restoring term in the perturbation equation is probably entirely thermodynamic in origin. If, as seems reasonable, the ideal gas law applies approximately to dark matter, the speed of sound will be

$$c_{\rm s} = \sqrt{\frac{\partial P}{\partial \rho}} \approx \sqrt{\frac{k_{\rm B}T}{\bar{m}}}\,, \tag{17.15}$$

where $\bar{m}$ is the mean particle mass, usually presumed to be ~ 100 baryon masses; so $c_{\rm s} \approx 500$ m/s for dark matter at photon decoupling. Since $T \propto a^{-1}$ (Equation (11.1)) the restoring term is proportional to $(c_{\rm s}/L_0)^2 \propto a^{-3}$, the same a-dependence as with the driving term; the ratio of restoring to driving terms will be constant for dark matter. The driving term dominates over the restoring term for dark matter perturbation masses of cosmological interest, so the perturbation equation for dark matter may safely be approximated by

$$\ddot{\delta} + 2H\dot{\delta} - \frac{3H_0^2\Omega_{\rm m,0}}{2}\frac{\delta}{a^3} \approx 0. \tag{17.16}$$

Late in the plasma era, when matter dominates but the Universe is still effectively flat, $H_0^2\Omega_{m,0}/a^3 = H^2\Omega_m \approx (2/3t)^2$ and the above equation reduces to

$$\ddot{\delta} + \frac{4}{3}\frac{\dot{\delta}}{t} - \frac{2}{3}\frac{\delta}{t^2} \approx 0\,, \tag{17.17}$$

for which the growing solution is $\delta(t) \propto t^{2/3}$. This is approximately the growth behavior for dark matter perturbation amplitudes after entering the horizon and after matter–radiation equality.[8]

Because of the relatively small sound speed, all dark matter perturbations of cosmologically interesting sizes can grow in density contrast during the later parts of the plasma era: the dark matter Jeans length remains within the horizon (see Figure 17.2), and smaller than the diameter of cosmological structures, for all but the earliest times in the plasma era. At photon decoupling the dark matter co-moving Jeans length is $L_0 \approx 2.3$ kpc and the Jeans mass is $\sim 150\ M_\odot$.

Coupled baryonic and dark matter

Gravitational driving of perturbation growth is proportional to the total gravitating mass, so the perturbation equation is properly written as two coupled equations

$$\ddot{\delta}_b + 2H\,\dot{\delta}_b = \left(\frac{3H_0^2\Omega_{m,0}}{2}\frac{\delta_b + \delta_d}{a^3}\right) - \frac{4\pi^2 c_s^2}{L_0^2}\frac{\delta_b}{a^2}\,, \tag{17.18}$$

$$\ddot{\delta}_d + 2H\,\dot{\delta}_d = \frac{3H_0^2\Omega_{m,0}}{2}\frac{\delta_b + \delta_d}{a^3} - \left(\frac{4\pi^2 c_s^2}{L_0^2}\frac{\delta_d}{a^2}\right)\,, \tag{17.19}$$

where subscripts b and d denote baryonic and dark matter perturbations, respectively; and the terms within parentheses are minor ones. The simultaneous solution of these two equations expresses the gravitational coupling between the two types of matter and results in perturbation growths that modify those of the two types from their independent behaviors.

The evolution of mass density perturbations for both dark and baryonic matter is illustrated schematically in Figure 17.3, which portrays integrations of the coupled perturbation equations for astronomically realistic perturbation amplitudes and diameters. As expected, the larger diameter perturbation (dashed curves) enters the horizon later and thus produces a larger (on average) baryonic amplitude at recombination, but also a smaller dark matter amplitude. The baryonic perturbations are the sources of anisotropies in the CMB radiation emergent from the plasma upon photon decoupling when electrons combine onto positive ions (next section), and the essentially invisible, and relatively large

[8] See Section 4.1.6 of Mo *et al.* (2010) for equivalent expressions during radiation dominance.

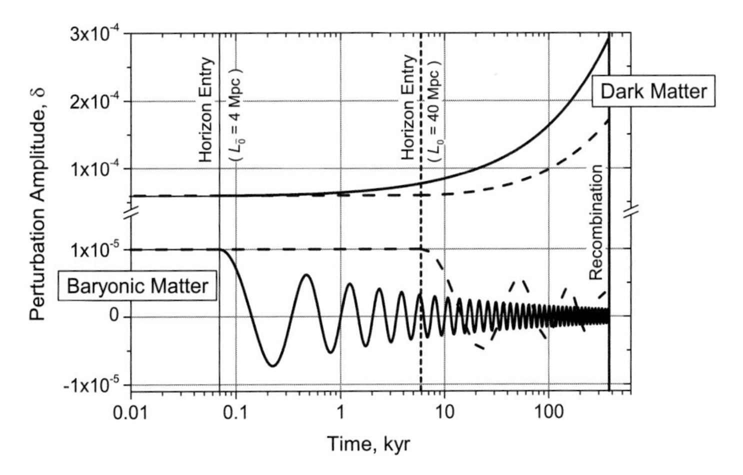

Figure 17.3 Schematic behavior of evolution of gravitationally coupled baryonic (lower panel) and dark (upper panel) matter perturbations. Solid curves: $L_0 = 4$ Mpc. Dashed curves: $L_0 = 40$ Mpc. Note: perturbation amplitude scale change between baryonic and dark matter plots.

amplitude, dark matter perturbations are probably the seeds for galaxy formation (next chapter).

17.2 Recombination

At $t \approx 375,000$ years, electrons combine onto positive hydrogen ions and thus matter effectively decouples from radiation so that baryonic matter becomes free to gravitationally collapse into denser structures. This marks the nominal end of the Plasma Era and the beginning of the era of galaxy formation.

Ionization and recombination of hydrogen may be represented by

$$P + e^- \rightleftarrows H + \gamma \ ,$$

where by H is meant the neutral hydrogen atom, while P (proton) denotes the ionized atom. The ionization energy of hydrogen is $\chi \approx 13.6$ eV, corresponding to the mean photon energy in black-body radiation of temperature $\sim 50,000$ K. Hydrogen can be ionized by radiation fields of much lower temperatures than this if there are sufficient photons so that ionization rates exceed those of electron capture onto ionized atoms; the ionization status of a gas thus depends upon its temperature *and* its density of both photons and free electrons – or, more precisely, on the number of energetic photons per atom or electron. High electron densities or pressures encourage electron captures, and drive ionization rates down, as the electrons are effectively pushed onto the ions; while high numbers of energetic photons increase ionization rates. Thus, in the photospheres of main sequence stars where electron number densities are relatively high, hydrogen ionization fractions of ~ 0.5 occur at temperatures near $\sim$ 10,000 K, and the ionization fraction in the Sun's photosphere, for which $T \sim 6000$ K, is small but measurable.[9]

In the Plasma Era, the number of photons per atom (or per electron) is $\sim 10^9$, so that the gas remains ionized to even lower temperatures than in stars. In a static environment the ionization fraction is given by the simple statistical equilibrium condition known as the Saha Equation,[10] which relates ionization to temperature and electron number density. But this equilibrium relation does not strictly apply to conditions during recombination when number densities and temperatures are changing rapidly, so the progress of recombination must be studied in terms of reaction rates, a much more complicated calculation. In summary: the hydrogen ionization fraction falls steadily as the Universe expands from $z \sim 1600$ to ~ 1100, when so few free electrons are left that photons are effectively decoupled from ions and electrons and are free to stream through the Universe unimpeded.[11] Strictly speaking, there is no 'recombination time': the process takes place gradually between $t \sim 200$ and ~ 375 kyr, but the photon scattering rate only falls significantly when electrons become very thin on the ground, so the time of last scattering – or of photon decoupling – is spread over a smaller interval and is conventionally taken as $t_{ls} \approx 375$ kyr at a redshift of $z_{ls} \approx 1090$. The photons released by recombination describe the 'last scattering surface' and are visible today (redshifted by z_{ls}) as the CMB radiation. That radiation carries with it the imprint of the matter perturbations created by inflation and modified by the intervening expansion of the Universe (see the previous section).

17.3 CMB anisotropies

The baryon density perturbations are made visible to us as temperature **anisotropies** (i.e., variations in temperature with viewing direction) in the last scattering surface: since $\delta \propto \Delta\varepsilon/\varepsilon = 4\Delta T/T$, the temperature variations across this surface map out the baryonic density perturbation distribution at photon decoupling time, revealed as CMB intensity variations at the observed frequency. The result is shown in Figure 17.4. The perturbations are small in density amplitude – on the order of 1 part in 10^5 – but are the only likely candidates for seeds of eventual galaxy formation and growth. By the Cosmological Principle, this is a representative slice of the temperature structure of the Universe's background radiation at photon decoupling. From the discussions of the previous sections, this pattern of temperature anomalies carries within it a good deal of information concerning the contents and evolution of the early Universe.

[9] See, e.g., Section 8.1 of Carroll and Ostlie (2007).
[10] Carroll and Ostlie (2007).
[11] See Section 9.3 of Ryden (2003) for a lucid and detailed description of the process of recombination.

Figure 17.4 Whole-sky CMB anisotropy map; temperature contrasts enhanced by $\sim 10^5\times$ (NASA/WMAP Science Team).

But the encoding of that information in the CMB anisotropies is not simple, being one of multiple overlapping perturbations of various sizes and amplitudes from which we must figure out the details of early expansion. In analyzing the CMB anisotropy structure we have the preceding theory of the growth of density perturbation as a guide, but only partially so: we do not know *a priori* the spectrum of primordial perturbations amplitudes and sizes, for instance; but must deduce them from the observed CMB anisotropy structure, guided by the informed speculation underlying the Harrison–Zel'dovich spectrum. One is reminded here of the analogy employed by the late Richard Feynman in describing spectroscopic analyses: it's like figuring out what's going on in a swimming pool occupied by swimmers and divers, only by studying the pattern of waves lapping against a small portion of one of the sides of the pool, and not looking up at the body of the pool.

We are helped in this analysis by regularities implicit in Figures 17.1 and 17.3, whose origin is a consequence of the Harrison–Zel'dovich spectrum. Since (according to this symmetry) all perturbations have the same density contrast upon entering the horizon and beginning their dynamical evolution, all baryonic perturbations of given linear size (and thus horizon entry time) will have the same phase upon photon decoupling, and thereby the same CMB temperature anisotropy amplitude. As a consequence, there is a good correlation between anisotropy angular diameter and amplitude, so that a mapping of observed temperature anisotropies vs. angular size should reveal features derivative of the physics of expansion.

17.3.1 Anisotropy spectrum

Problems of such complexity are best studied statistically: how much of the observed anisotropy distribution is in the form of perturbations of given size and amplitude? The appropriate statistical tool is the spherical harmonics analog of

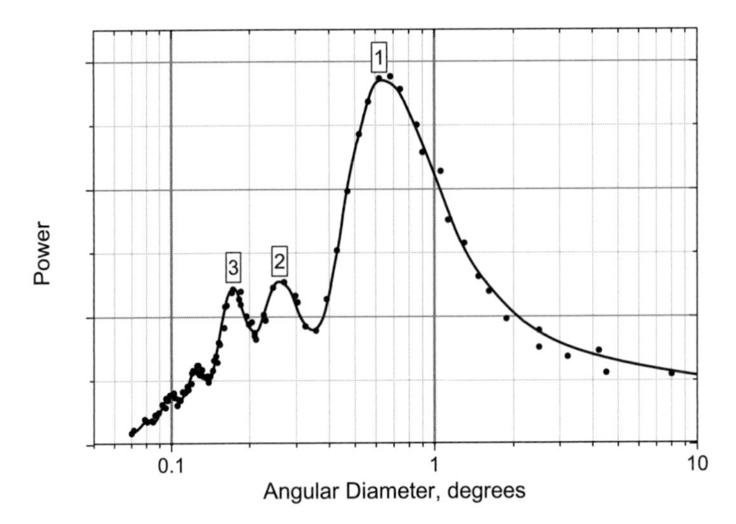

Figure 17.5 The observed CBR anisotropy power spectrum in angular units. Points are representative observations; the solid line is a theoretical fit from the CCM with dark matter (see text).

a Fourier power spectrum of the observed anisotropy sizes and amplitudes. The result is conventionally summarized as a graphical representation of the amount of structure found on different scales.[12] For the observed CMB anisotropies the result from numerous data sources is shown in Figure 17.5.[13] There is a principal peak (labelled 1) at an angular scale of ~0.6 degrees, with subsidiary peaks at smaller scales (and smaller power amplitudes).

The origin of this structure is basically acoustic. The largest baryonic perturbation that could reach its maximum density would have a diameter corresponding to the sound horizon distance at decoupling time, the maximum distance a sound (pressure) wave could have travelled in that time. From Equation (10.14) this distance is

$$d_{\mathrm{SH}}(t) = a(t) \int_0^t \frac{c_{\mathrm{s}}(\tau)}{a(\tau)} d\tau \,. \tag{17.20}$$

For $a(t)$ of the CCM and $c_{\mathrm{s}}(t)$ given by Equation (17.13), this integral is $\approx$ 140 kpc at photon decoupling time. This is approximately the largest feature

[12] See, e.g., Chapter 6 of Peter and Uzan (2009) for a detailed discussion of the rather difficult physics underlying such power spectrum analyses, and their appendix B.2 for a summary of the properties of spherical harmonics.

[13] This figure is the mirror image of that normally published, for which the independent variable is the spherical harmonic multipole moment l, related approximately to angular size by $l \sim 140/\theta$ for θ in degrees. Data and theoretical fit from D. Scott and G. F. Smoot, Section 23 of J. Beringer *et al.*, Review of Particle Physics, (2012), *Phys. Rev. D.*, **86**.

observable on the last scattering surface and corresponds to a baryonic perturbation that has just achieved its maximum density at photon decoupling, with an oscillation 'phase' corresponding to $1/4$ of the period so that the sound horizon is $1/4$ of the perturbation wavelength. From Equations (9.44) and (9.85) for the CCM at decoupling redshift $z \approx 1090$, this corresponds to an angular width on the observed last scattering surface of $\theta \approx 0.63$ degrees, which is approximately what is observed (see Figure 17.5).

Smaller (in diameter) perturbations will have relatively large density amplitudes at photon decoupling – and thus be apparent on the last scattering surface – if their oscillation phases there correspond to oscillation extrema; i.e., if they are (approximate) overtones of the principal perturbation represented by the sound horizon. Since the oscillation frequency is inversely proportional to perturbation diameter, the result is a pattern of hot and cold patches on the CMB of angular sizes roughly equal to odd integer fractions $(3, 5, \ldots)$ of 0.63 degrees, approximately what is observed (see Figure 17.5).

That higher-order (smaller diameter) acoustic peaks are generally lower in amplitude than their predecessors is due to two effects. First, smaller diameter perturbations enter the horizon earlier (Figure 17.2) and suffer more oscillations (Figure 17.1), and thus are subject to more Hubble damping, than is the case with larger perturbations. Second, photons may leak out of the baryon/photon perturbations during contraction, reducing expansion pressure and the resulting density amplitudes.[14] The effect is greater for smaller perturbations, whose diameters would be smaller in comparison with the mean photon scattering length, hence the systematic reduction in acoustic peak amplitudes with order. As is apparent from Figure 17.5, perturbations of co-moving angular diameters much less than ~ 0.1 degrees are effectively damped out by photon decoupling time.

That acoustic peaks numbers $\boxed{2}$ and $\boxed{3}$ do not conform to this pattern is a consequence of dark matter. Dark matter feels no radiation pressure and so oscillates in perturbation amplitude entirely because of gravitational coupling to baryons. During contraction the dark matter in a density perturbation acts in concert with baryons, but during expansion the two forms of matter oppose each other: dark matter wants to continue contracting while baryonic matter is being pushed apart by radiation pressure. Acoustic peak $\boxed{2}$, immediately following the maximum contraction of the primary peak, corresponds to maximum expansion and thus is of a lower amplitude because of dark matter; while acoustic peak $\boxed{3}$ is larger than it would have been without the influence of dark matter. Even numbered baryonic acoustic peaks are systematically lower, and odd numbered ones higher, than would be the case without the influence of dark matter.

[14] A process known as *Silk damping*, after the cosmologist Joseph Silk who first explained the phenomenon.

Sachs–Wolfe effect

Mass perturbations larger than the sound horizon at decoupling time cannot appreciably evolve by that time, but their effect can still be seen in the CMB anisotropy spectrum. The density enhancement in a mass perturbation creates a gravitational density well: photons emerging from that well are red-shifted so that a CMB temperature anisotropy is created. The result is structure in the CMB spectrum on scales exceeding the CMB anisotropy principal peak; this is visible to some extent in the high-diameter tail shown in Figure 17.5. The effect was first studied by Sachs and Wolf[15] who derived the qualitative connection

$$\frac{\delta T}{T} = \frac{1}{3}\frac{\delta\Phi}{c^2}\,, \tag{17.21}$$

where $\delta\Phi$ is the gravitational potential change induced by the perturbation.

A somewhat related effect arises from gravitational potential changes during the Universe's expansion following decoupling. In this *integrated Sachs–Wolfe* effect, photons falling into a gravitational well are initially blue-shifted, then later red-shifted as they emerge on the other side. If the well's depth decreases during photon transit across it, the net result will be a blue-shift whose magnitude is[16]

$$\left(\frac{\delta T}{T}\right)_{\vec{r}} = \frac{2}{c^2}\int_{t_e}^{t_0} \frac{\partial\Phi\left(\vec{r},t\right)}{\partial t}dt\,. \tag{17.22}$$

Since the acceleration induced by dark energy is the most likely mechanism for reducing gravitational potentials in an expanding Universe, this observable (in principle) feature is a promising venue for studying the properties of dark energy.

Polarization

Quantum fluctuations during Inflation would lead to large irregularities in the underlying gravitational field which would induce gravitational radiation, the residues of which should propagate into the current time. These should be observable on large scales, as both polarization and anisotropies in the CMB. The expected polarization signals seem especially promising for detection of such waves, since they would be a nearly unique product of Inflation.[17] But the polarization amplitudes would be at least an order of magnitude less than their temperature amplitudes, and thus very difficult to detect. As of this writing, no such signals have been unambiguously detected, although some experiments appear to be tantalizingly close to anticipated detection limits. Stay tuned.

[15] Sachs, R. K. and Wolfe, A. M. (1967), *Astrophys. J.* **147**, p. 73.

[16] Liddle and Lyth (2000), Section 6.2.2.

[17] See, e.g., Sections 6.5 and 6.6 of Liddle and Lyth (2000) for further discussion of gravitational waves and curvature attendant upon inflation.

The importance of studying the consequences of inflation, and particularly of the gravitational waves produced therein, is that such studies provide a link between gravitation and quantum fields. These two field theories underlie all of modern physics but are quite independent: while we think they are connected at a deep level, we currently have no useful theory relating them. A theory joining them is, arguably, the Holy Grail of modern physics.

17.3.2 CMB anisotropy diagnostics

The overall properties of the CMB anisotropies spectrum, as summarized in part in Figure 17.5, constitute a striking confirmation of the CCM and associated particle physics. The CMB anisotropy measurements place constraints on cosmological parameters that severely limit the possibilities of significant departures from the CCM values. While no one CMB anisotropy feature defines any one CCM parameter – there is substantial degeneracy in the mapping of anisotropy features onto CCM parameters – two features of the anisotropy spectrum are particularly useful in nailing down CCM expansion.

Curvature and dark energy The physical diameter corresponding to the anisotropy principal peak ($\boxed{1}$ in Figure 17.5) is sensitive to the (relatively well-known) matter/radiation densities but not to curvature or dark energy, since CMB formation takes place in a dynamically flat environment at a time when dark energy is an insignificant gravitational component. But the *projected* angular diameter of this feature – as observed today – involves propagation of light rays through the $0 \lesssim z \lesssim 1090$ Universe where curvature may contribute: refer to Figure 15.5 as an example. For given physical diameter the observed angular diameter will be greater in a positively curved geometry, and less in a negative one. The observed ≈ 0.63 degrees is a good match to the predicted $L \approx 140$ kpc in a flat geometry; this match arguably constitutes the most direct evidence that the Universe is flat, or nearly so. In turn, this is indirect – but compelling – evidence for dark energy, since matter and gravitation by themselves almost certainly are insufficient to close the geometry.

Dark matter The nearly equal power in the first two power spectrum overtones – peaks $\boxed{2}$ and $\boxed{3}$ in Figure 17.5 – can be accounted for by a dark matter density that is ~ 6 times that of baryonic matter, but seemingly in no other way. The importance of this evidence is that it is non-gravitational in origin, in the sense that it probably cannot be explained by alternative theories of gravity. In particular, the proposed Modified Newtonian Dynamics (MOND) theory (Section 12.6) would seem to be refuted by this feature.

Problem

1. Find a second solution (in addition to $\delta \propto t^{2/3}$) to the dark matter perturbation equation (17.17). Why is it not the preferred solution for our purposes?

18

Galaxy Era

Immediately following recombination, the Universe effectively consisted of a gas of neutral hydrogen and helium atoms immersed in dark matter, within a sea of photons. The baryonic matter content of the Universe was thin and nearly uniform, with density irregularities of at most $\sim$ 1 part in 10^5. The Universe's radiation content – which would become the CMB radiation – was that of a blackbody of temperature $\sim$ 3000 K, and thus mostly in near infrared wavelengths. Had there been eyes to see with at that time the Universe would have resembled a thin, hazy cloud illuminated with a photon spectrum resembling that of an ordinary incandescent light bulb. As the Universe expanded it became thinner and dark, with CMB photons redshifted into the infrared.

By the age of $\sim 1/2$ billion years – less than 4% its current age – the Universe had become clumpy as the first stars and rudimentary galaxies formed, producing radiation that re-ionized the intergalactic medium and endowed the Universe with visible light. A few billion years later the Universe rather resembled its current form, with gravitationally mediated structures spanning a wide range of spatial and mass scales and a radiation content that included UV and visible radiation from stars. How that coagulation of baryonic matter took place is a matter of great interest to cosmologists, and the subject of much current research.

18.1 Structure formation

A major clue to structure formation in the Universe comes from the analysis of perturbation growth during the Plasma Era. Immediately following photon decoupling the speed of sound in baryonic matter drops from $\sim c/2$ to $\sqrt{k_B T/\bar{m}_B} \approx 5$ km/sec and the baryonic Jeans mass falls to $\sim 10^6$ solar masses, the size of a very large globular cluster or a very small galaxy; so that even small cosmological structures can condense further. Initially, further baryonic perturbation amplitude growth proceeds as with dark matter: $\delta_b(t) \propto t^{2/3}$ (Section 17.1.2) which, since matter now dominates the expansion and $a \propto t^{2/3}$, implies $\dot{\delta}_b/\delta_b \approx \dot{a}/a$. Since a has increased only about 1000-fold since decoupling, and the largest observed temperature irregularities in the CMB are $(\Delta T/T) \lesssim 10^{-5}$, the largest purely baryonic density perturbations at the current time would be $\delta_b\,(t_0) \approx 10^{-2}$, many orders of magnitude less than what is observed in stars and galaxies. What is evidently needed to produce our current Universe is an accelerator for baryonic structure growth. That mechanism appears to be providentially provided by dark matter.

Dark matter density perturbations in the Galaxy Era would initially be much larger in amplitude than those of baryons: from the example shown in Figure 17.3 we can reasonably expect dark matter density contrasts to be several orders of magnitude larger than those of baryons at the time of photon decoupling, and that the 'small perturbation' approximation for dark matter would become invalid not long after photon decoupling (if not before). It is thus reasonable to suppose that large dark matter density contrasts could grow early in the galaxy era and have served as gravitational attractors for baryonic matter condensations, a supposition supported by observations of massive dark matter 'halos' about galaxies such as the Milky Way. Presumably, relatively low-contrast baryonic matter concentrations would fall into the potential wells of the much more massive dark matter concentrations and there condense or combine to produce the baryonic structure observed in the form of galaxies and galaxy clusters. The story of galaxy and larger structure formation in the early Universe is thus largely one of dark matter dynamics.

Such a scenario has become the basis of the 'cold dark matter' model of structure formation in the galaxy era. The 'cold' in this phrase should really be read as 'non-relativistic', the point being that, unlike with photons and neutrinos, non-relativistic matter can gravitationally contract into enhanced density structures. But note that the presumed dark nature of this matter – reflecting very weak or non-existent coupling to electromagnetism – implies an inability of the gas to cool by radiation, so that dark matter cannot condense beyond the point of thermal pressure support in a relatively hot gas.[1] It is thus not surprising that the dark matter associated with most galaxies appears to be in the form of very extended, massive halos of relatively low mass densities (and, presumably, high kinetic temperatures).

18.2 Dark Ages

The era following recombination was without sources of illumination (other than the diffuse, and long-wave, CMB radiation) and is known as the 'Dark Ages',

[1] Dark matter could also cool by mechanical interactions with cooler baryonic matter, but this mechanism is much less efficient than radiation.

for obvious reasons. The Dark Ages came to an end with the formation of the first stars. Stars probably did not form independently, but within the relatively dense environments of galaxies or proto-galaxies. Their formation illustrates a generic issue with gravitational structure formation: the dissipation of gravitational and thermal energy necessary for substantial density enhancements. In the current Universe, star formation occurs when a gas cloud exceeds its Jeans mass or length, usually by cooling or compression. As the cloud gravitationally collapses its density increases and thus its Jeans mass decreases, and the cloud fragments into – eventually – proto-stars. The fragmentation process stops when fragments can no longer dissipate thermal energy generated by gravitational compression, so that their internal temperatures and Jeans masses rise. Stellar masses are thus set, in part, by radiative processes mediated by gas opacities. Put succinctly: stellar masses are a consequence of cooling mechanisms; inefficient cooling of the collapsing cloud leads to relatively massive stars.

As it happens, pure hydrogen/helium gas mixtures cannot cool themselves as efficiently as can gases with heavier elements intermixed, so that the first stars to form from only hydrogen and helium – 'Population III stars' in the parlance – were probably massive ones, in excess of $\sim$ 100 solar masses. Being massive these objects would also be short-lived, hot, and the source of copious amounts of UV radiation capable of ionizing the neutral hydrogen gas constituting the bulk of matter in the Universe at that time. This ionization of hydrogen following formation of the first stars is known as 'reionization': the current best observational evidence for it is the 'Gunn–Peterson Trough', a hole in the Lyman-alpha forest blueward of the red-shifted line in distant quasars attributed to absorption by neutral hydrogen atoms between us and the illuminating quasar. The trough is not present in quasars with redshifts less than about 6.3, an observation generally interpreted as indicative of re-ionization between $\sim$ 0.5 and 1 Gyr.

18.3 Galaxies

The formation and early evolution of galaxies is not an easy subject for study, confused as it is by radiative and mechanical feedback to the formative gas from such structures: shock waves from supernovae, ionizing radiation from massive stars, etc.; what one wag has called 'gastro-astronomy'. The study of galaxy formation is hampered by remoteness in time: unlike stars that are forming today, the formation and much of the evolutionary history of galaxies took place when the Universe was much younger, so that the early histories of galaxies cannot easily be observed. The history of galaxy formation and early evolution is further obscured by the evident fact that most galaxies do not evolve in isolation, so that their current features reflect a history of galaxy mergers and interactions – also mostly in the remote past. Overall, galaxy formation and evolution is a large and complex study, worthy in itself of several books of this length; see, e.g., Loeb and Furlanetto (2013); Longair (2008); Mo, van den Bosch, and White (2010); Liddle and Lyth (2000). In this chapter we summarize some of the main features of current studies of galaxy formation and evolution, without the more esoteric details.

Several lines of evidence and reasoning point to a formation scenario in which galaxies form 'from the bottom up'. Observations of very distant galaxies ($z \gtrsim 4$) show objects that are smaller, less massive, and less regular in appearance than are typical galaxies in the current Universe.[2] This is to be expected from the density perturbation population characteristics discussed in the previous chapter. As illustrated in Figure 17.3, smaller (in diameter) perturbations enter the horizon earlier and thus experience more enhancement of their dark matter densities than do their larger brethren, and so are more likely to form the seeds of galaxy formation in the post-recombination era. These small galaxies probably grow to their present size by a combination of mergers and accretion, complex processes that have led to the current variations in galaxy sizes and types.

Galaxy mergers are not unknown in the current Universe, and must have been common in the denser, earlier Universe. From statistics of observed merging galaxies at different redshifts it seems likely that most large galaxies ($M \gtrsim 10^{10}$ $M_\odot$) have undergone several 'major' mergers in the past, each one approximately doubling the galaxy mass.[3] Minor mergers, as with dwarf galaxies, are clearly taking place today: the Milky Way Galaxy is probably in the process of ingesting several such neighbors, including the well-observed Sagittarius Dwarf Spheroidal galaxy that appears to be spiralling into the Galaxy. Less visible, but possibly equally important, is the transfer between galaxies and the intergalactic medium. Considering that galaxies such as the Milky Way have only $\sim$ 15% of their baryonic disk mass in the form of interstellar material, and that they are forming $\sim$ 1 star per year from that gas, it seems likely that replacement gas has been needed to preserve the Galaxy's stellar population statistics over the course of its later history.

Galaxies appear to be the main, perhaps the only, venue for star formation in the Universe. Mean star formation rates in the Universe appear to have peaked at $t \sim 2 - 3$ Gyr and have been gradually declining ever since.[4] By $\sim$ 7 Gyr galaxy formation must have largely ceased due to universal acceleration (which will increase the Hubble drag on structure formation beyond that otherwise expected). About half the stars in the universe today had already formed by the time the

[2] See, e.g., Figures 7–10 of Beckwith *et al.* (2006), *Astron. J.* **132**, p. 1729. Their Figure 10 is reproduced as Figure 19.5b in Mo *et al.* (2010).

[3] C. Conselice, *Physics Today*, August 2011 p. 68.

[4] See, e.g., Figure 2.35 of Mo *et al.* (2010).

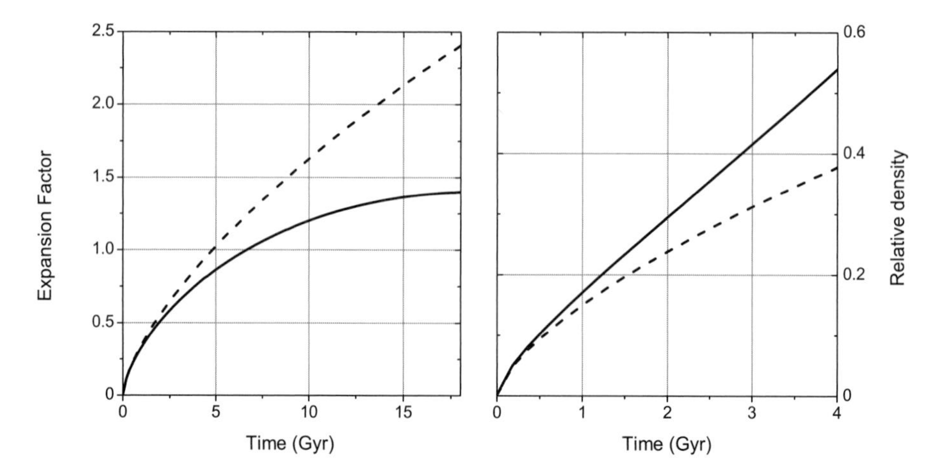

Figure 18.1 Non-linear perturbation growth following recombination: solid curve, density perturbation; dashed curve, unperturbed background. Left panel: growth of expansion functions a. Right panel: growth of relative density perturbations.

Universe was half its current age, and most of the stars that will ever form have already done so.[5]

18.4 Large-scale structure

Following recombination and photon decoupling, overly dense portions of the Universe may be treated as independent 'sub-universes' that may be modelled as structures whose evolution is imposed onto that of the background Universe. The appropriate model would be a closed, matter-only one whose expansion would be governed by the cycloidal Equations (12.4). The expansion function in these regions would increase more slowly than in the Universe at large, as shown in Figure 18.1 (left panel); so that the density contrast would eventually increase more rapidly than $\propto a$, as shown also in Figure 18.1 (right panel). The result is that density perturbations of sufficiently large amplitude at photon decoupling time – almost certainly composed entirely or largely of dark matter – could have grown more rapidly than they did in the pre-recombination era.[6]

The statistical distribution of structure sizes – e.g., how many small clusters vs. how many large ones – can be predicted from models of evolution of the Universe's matter and energy densities, and their incorporation into condensed

[5] C. Conselice, *Physics Today*, August 2011, p. 69.

[6] See Sections 11.4, 16.1 of Longair (2000), or Section 5.1 of Mo *et al.* (2010), for detailed analysis of this 'Top Hat' collapse modelling.

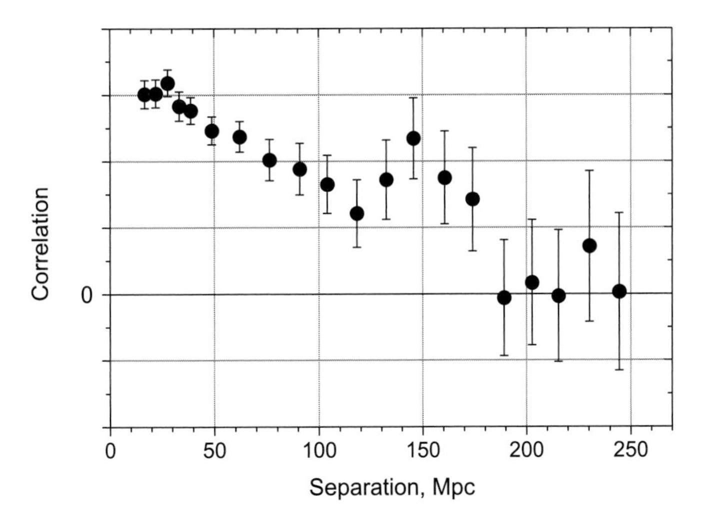

Figure 18.2 Observed two-point galaxy correlations as a function of current proper separation (data adapted from Bennett 2006, *Nature* **440**, p. 1126).

structures. The current standard is the 'ΛCDM' model – cold dark matter with Cosmological Constant. The 'cold' here refers to non-relativistic matter (as opposed to, say, massless neutrinos). This model has rather successfully reproduced the largest cosmological structure features, as shown in Figure 18.2. The large 'bump' near 150 Mpc is apparently a fossil remnant of the principal peak in the CMB anisotropy spectrum (Figure 17.5), reflective of the sound horizon at photon decoupling time. Such features in the current Universe are commonly referred to as baryonic acoustic oscillations (BAO), and constitute another bit of evidence for the existence of dark matter (if any were needed).

In reality, density perturbations are unlikely to be spherically symmetric as implicitly assumed in the analysis thus far. As a first step toward more realism, Zel'dovich and others modelled the gravitational evolution of triaxial ellipsoids and showed that the fastest contraction took place along the shortest axis, so that ellipsoids first became sheets, then linear structures, then (possibly) compact objects.[7] The result is a filamentary, cellular structure commonly called the 'cosmic web' (for obvious reasons) and illustrated in Figure 18.3. Observations of large-scale structure in the Universe – clusters and superclusters – suggest that this sort of dark matter webbing underlies such structure.

On large scales the current Universe's structure appears to be cellular, with galaxy clusters strung along cell walls encompassing relatively empty spaces tens of Mpc in diameter. This is shown clearly in the SDSS (Sloan Digital Sky Survey) observations of galaxies and clusters plotted in Figure 18.4 and is quite consistent with computer simulations of structure formation dominated by non-relativistic

[7] Zel'dovich, Y. (1970), *Astron. Astrophys.* **5**, p. 84.

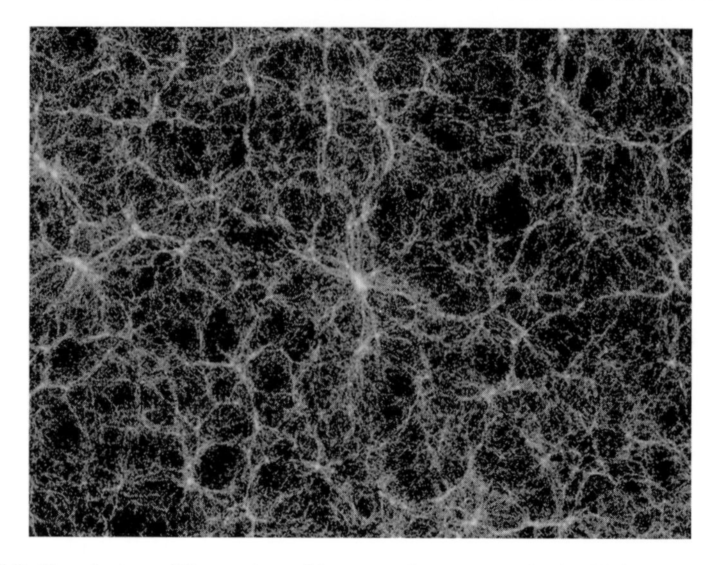

Figure 18.3 Simulation of formation of large-scale structure in the Universe (Millennium Simulation).

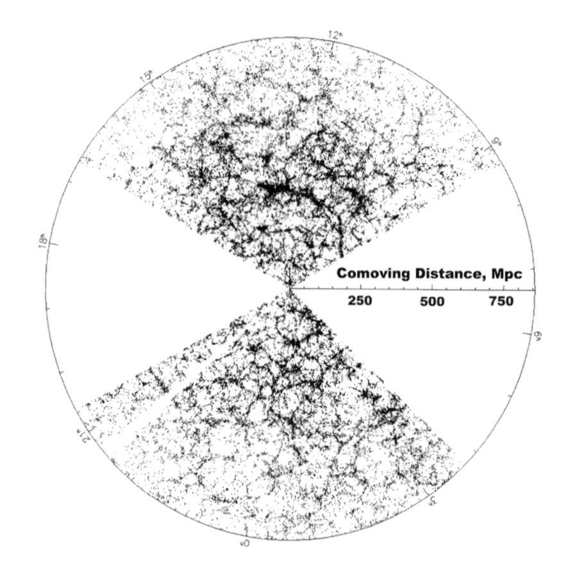

Figure 18.4 Spatial distribution of galaxies and clusters in a wedge 2.5 degrees wide in declination, out to about 900 Mpc (adopted from SDSS/M. Blanton data).

matter in an expanding framework, even if the finer details of galaxy formation are yet to be understood. In effect, the current large-scale structure of the Universe reflects the distribution of CMB anisotropies from which structure has arisen: compare this figure with Figure 17.4. Paraphrasing the late Carl Sagan: this is what gravitation can do, given thirteen billion years.

18.5 The future ...

Thirteen+ billion years after it all began with the formation of quantum fluctuations in the inflationary era, gravitation and expansion have changed the nature of the Universe from that of a nearly featureless plasma into a structured arrangement of matter and radiation in which – paraphrasing Martin Rees (1995) – the primordial inhomogeneities have been amplified, and temperature gradients steepened, to the point that the complexity of which we are a part has come into being. It remains for cosmologists and others to explain why this has happened in the way it has: why, for instance, the structure of the Virgo cluster is one of discrete, interacting galaxies and not just a homogeneous collection of $\sim 10^{14}$ stars; and how chemically dynamic structures like our Milky Way Galaxy have come into being and lasted as long as they have. Some aspects of this and similar deep issues are discussed in the final chapter.

The Universe is currently expanding at nearly an exponential rate so that dynamical collapse times for most cosmological structures are longer than the Hubble expansion time, or nearly so; and certainly will be so in the immediate future. New gravitational structures cannot form in such circumstances, and existing ones are threatened with disruption in the near future – i.e., within a few tens of Gyrs. The event horizon in the CCM is ≈ 4.9 Gpc and will remain nearly constant for the foreseeable future, a distance that will be exceeded by the dimensions of all current structures in the Universe in the fullness of time. Thus, after ~ 100 Gyr the Milky Way Galaxy, or what will be left of it at that time, will fill its event horizon and will effectively be cut off from the rest of the Universe. By ~ 1000 Gyr the event horizon will be filled by just one wavelength of the (by then greatly redshifted) CMB radiation. Of course, all this is subject to revision if the dark energy currently driving universal expansion should change or disappear, much as happened with the energy field driving Inflation. Lacking a good theoretical framework for studying such matters, we cannot confidently predict the future.

It nonetheless appears that we live in an interesting and unusual time in the Universe's history. Later generation stars with solid planets like the Sun could not have formed until the Universe was several Gyrs old, and could not have harbored evolved life until some time after that. Many Gyr after the current time, it will probably not be possible for such things to come into being in the exponentially expanding and thinning Universe. It is fair to say that the future Universe may be very different from the one we live in, raising interesting questions that border on being more philosophical than scientific in that they call into question the issue of the meaning of human life.

19

Afterword: the new modern cosmology

> To see a World in a Grain of Sand
> And a Heaven in a Wild Flower,
> Hold Infinity in the palm of your hand
> And Eternity in an hour.
>
> William Blake,
> *Auguries of Innocence*

Consider now the current state of our understanding of the expanding Universe. We appear to have a good understanding of the fundamental physics underlying its expansion, of the origin of structure in the current Universe, and of its detailed history and likely future. Our current cosmological world-view is remarkable not only for its scope but also for its coherence: everything seems to hang together in a nearly seamless picture that purports to explain nearly all of the Universe's large-scale structure and evolution, from shortly after the moment of creation to the current time, and on scales dwarfing anything else in our experience or intellectual musings.

Consider, for instance:

- the dark matter originally proposed to explain intracluster galaxy velocities and individual galaxy rotation rates is just what is needed to explain details of the CMB anisotropy spectrum; the formation of baryonic galaxies embedded in massive, dark halos; and the large-scale structure of the Universe in the form of galaxy clusters and super-clusters;
- the dark energy originally inferred from SN Ia Hubble relations also accounts for the flatness of the Universe, its current age, details of the CMB anisotropy spectrum, and the Universe's large-scale structure;
- universal Inflation connects theories of quantum fields to gravitation in ways that help explain the large-scale features of the Universe (flatness, homogeneity), and provides the seeds for eventual structure formation;

- the universal abundance of helium is understood to be a fossil remnant of Baryogenesis in the very early Universe, leading to the large photon/baryon ratio and large entropy in the current Universe.

All in all, modern theoretical cosmology is a remarkable intellectual achievement, especially so when compared with the state of our knowledge as recently as 50 years ago – within the lifetime of many cosmologists working today – when the nature of quasars was unknown, the CMB had yet to be observed (let alone its structure), the Steady State model was a viable alternative to the Big Bang, and such things as universal acceleration and Inflation were not dreamed of. Cosmologists can justifiably be proud of the advances made in their science in the past several decades.

But our current theory of cosmology, as impressive as it seems at first sight, is incomplete at a worrisome level. Particularly troublesome is our lack of understanding of the nature and origin of dark matter and dark energy, which together encompass $\sim 95\%$ of the Universe's energy density; and of the fundamental physics underlying the Universe's early history, including Inflation. Among the ways in which uncertainties in these matters reveal themselves in cosmological details are the values of seemingly universal constants that are not constrained by current theory. In particular:

- The universal structure constant, $\psi \approx 10^{-5}$, which limits the amplitudes of primordial density fluctuations created during inflation as revealed in the CMB anisotropy amplitudes, and carried into current large-scale structure in the form of baryonic acoustic oscillations. This same number also seemingly characterizes the internal structure of galaxy clusters, in the sense that the typical intracluster kinetic energies of galaxies are $\sim 10^{-5}$ of their rest energy.
- The photon/baryon number ratio, $\eta^{-1} \approx 2 \times 10^9$, which presumably arose from a small particle/anti-particle asymmetry during Baryogenesis so that the large number of photons arose from mutual annihilations, with only a small fraction of hadrons and leptons escaping destruction and eventually forming the observed baryonic structures in the current Universe.
- The dark energy density of $\varepsilon_{de} \approx 6 \times 10^{-10}$ J/m^3, which is absurdly lower than predicted by current quantum field theory (' ...probably the worst theoretical prediction in the history of physics!').[1]
- The very small curvature of the Universe, quite possibly exactly zero; reflecting a fine balance between the Universe's gravitating contents and its rate of expansion.

[1] Hobson and Efstathiou (2006), p. 187.

The result, in Lee Smolin's words,[2] is "... [a] preposterous universe with its seemingly inexplicable fine tuning of parameters, both elementary and cosmological ...". Noteworthy in this regard is the observation that properties of the current Universe – including those needed for life to have arisen – are sensitive to the values of these constants. Of course, the same could be said for constants not confined to cosmology, such as the gravitational constant G, the neutron/proton mass difference Q, the atomic fine structure constant α; and several others.[3] We cannot confidently claim that we understand how our Universe came to be as it is, without understanding why such fundamental things have the values they do.

If pressed on the matter, most theoretical cosmologists would probably confess to the hope that resolution of these issues would only augment, and not necessitate major changes in, current cosmological theory. But the history of science should not encourage them in this belief.

A little over a hundred years ago, in the late nineteenth century, all of basic physics seemed to be understood, with robust and seemingly complete theories in mechanics, electromagnetism, and thermodynamics. James Clerk Maxwell opined that there would be nothing more for future generations of physicists to do but add more decimal points to the values of fundamental constants. Of course there were a few worrisome but hopefully minor matters yet to be explained, principal among them being the radiative spectrum of black bodies, the nature of the luminiferous aether seemingly needed for propagation of light, and the mysterious rays emanating from electrical discharges and some minerals.

These discrepancies turned out to be not minor at all, but harbingers of major changes required at the roots of physical theory. Within the first decade of the twentieth century Planck and Einstein showed that resolution of the first two of these inconsistencies required major overhauls of all of physics, leading to quantum mechanics and relativity. Within twenty years these new theories had completely disrupted the world of physicists, setting the stage for a new generation to build what we now call 'modern physics' and effecting one of the great revolutions in the history of science.[4]

Which is not to say that a similar scientific revolution is likely to occur in cosmology: predicting major intellectual change is a risky business at best, and no prudent author would do so in a textbook such as this. But the similarities between today's cosmology and late nineteenth-century physics are striking, and

[2] Unger and Smolin (2015) p. 355.

[3] See Martin Rees' popularization, *Just Six Numbers*, Basic Books (2000), for elaborations on these matters and attendant speculations on anthropic consequences.

[4] A social and scientific revolution hauntingly described in Russell McCormach's novel *Night Thoughts of a Classical Physicist*, Harvard University Press (1991).

should alert us to the possibility that major advances may follow from resolution of the incomplete sections of modern cosmology. Lurking in the background here, rather like Banquo's ghost at the feast, is the specter of quantum gravity, possibly *the* major unresolved issue in modern physics. It is entirely possible that a successful union of quantum field theory and gravitation – which, given the resources being devoted to its discovery, should occur within a current student's lifetime – may resolve all the unsettled basic issues in cosmology. Until then, the prudent cosmologist should be wary of confident conclusions and be prepared to change intellectual direction when and if new observations compel them to do so.

Part VI

Appendices

Appendix A

Differential geometry

A.1 Affine connection

The fundamental difficulty with doing mathematics in non-flat geometries is that the directions of unit vectors change with position, so that differentiation becomes complicated since changes in quantities over distance can be confused with changes in coordinate orientations. This is the source of, e.g., the complicated terms in gradients and divergences when using polar coordinates in flat geometries. An approach to dealing with such problems in curved geometries is one in which vector quantities are defined in 'planes' (lower dimensional surfaces) tangent to the curved surface; this turns out to be best described by a version of geometry characterized by quantities that are preserved under linear transformations and translations, and is equivalent to Euclidean geometry without the third and fourth postulates. The properties of such geometries were historically called *affinities*, and the geometries themselves have come to be called *affine geometries* and are considered to be special cases of the more general field of projective geometry.

The metric quantity needed to go from a curved surface to a tangent one is called a *connection*, for which there are many possibilities. The one most useful in the type of differential geometry used in GR is the **affine connection**, $\Gamma^{\alpha}_{\beta\gamma}$; loosely defined as the α-component change in the β-unit vector as one moves along the γ-axis. Symbolically (and employing the summation convention),

$$\frac{\partial \hat{\mathbf{e}}_{\gamma}}{\partial x^{\beta}} = \Gamma^{\alpha}_{\beta\gamma} \hat{\mathbf{e}}_{\alpha}, \tag{A.1}$$

where $\hat{\mathbf{e}}$ is the unit vector in the chosen coordinate system. This turns out to be a most useful quantity in differential geometry.

Metric formulation

For an arbitrary vector $\mathbf{v} = v^\mu \hat{\mathbf{e}}_\mu$,

$$\frac{d\mathbf{v}}{d\tau} = \frac{d}{d\tau}\left(v^\mu \hat{\mathbf{e}}_\mu\right) = \frac{dv^\mu}{d\tau}\hat{\mathbf{e}}_\mu + v^\mu \frac{d\hat{\mathbf{e}}_\mu}{d\tau} ,$$

for some scalar parameter τ. Now set $\mathbf{v} = d\mathbf{x}/d\tau$ so the above equation becomes

$$\begin{aligned}\frac{d^2\mathbf{x}}{d\tau^2} &= \frac{d^2x^\mu}{d\tau^2}\hat{\mathbf{e}}_\mu + \frac{dx^\mu}{d\tau}\frac{d\hat{\mathbf{e}}_\mu}{d\tau} ,\\ &= \frac{d^2x^\mu}{d\tau^2}\hat{\mathbf{e}}_\mu + \frac{dx^\mu}{d\tau}\frac{dx^\nu}{d\tau}\frac{\partial\hat{\mathbf{e}}_\mu}{\partial x^\nu} ,\\ &= \frac{d^2x^\mu}{d\tau^2}\hat{\mathbf{e}}_\mu + \frac{dx^\mu}{d\tau}\frac{dx^\nu}{d\tau}\Gamma^\alpha_{\mu\nu}\hat{\mathbf{e}}_\alpha ;\end{aligned}$$

the last step from Equation (A.1). Rationalizing dummy summation indices,

$$\frac{d^2\mathbf{x}}{d\tau^2} = \left[\frac{d^2x^\alpha}{d\tau^2} + \frac{dx^\mu}{d\tau}\frac{dx^\nu}{d\tau}\Gamma^\alpha_{\mu\nu}\right]\hat{\mathbf{e}}_\alpha .$$

Specifically: for a freely falling object in an LIRF where τ is the proper time, $d^2\mathbf{x}/d\tau^2 = 0$ so that

$$\frac{d^2x^\alpha}{d\tau^2} = -\frac{dx^\mu}{d\tau}\frac{dx^\nu}{d\tau}\Gamma^\alpha_{\mu\nu} \quad \text{(freely falling)} ,$$

for any index α. This is just the freely falling equations of motion derived in Section 5.3 (Equation (5.7)), where Γ was defined as (Equation (5.6))

$$\Gamma^\alpha_{\mu\nu} \equiv \frac{\partial x^\alpha}{\partial \xi^\lambda}\frac{\partial^2 \xi^\lambda}{\partial x^\nu \partial x^\mu} , \tag{A.2}$$

where ξ is a Cartesian coordinate system in flat space-time and $\{x\}$ is a freely falling system, defined in terms of ξ. It follows that the two expressions for the affine connection, Equations (A.1) and (A.2), are equivalent, and that the connection is symmetric in its two lower indices.

Neither of the coordinate expressions for Γ is convenient for most applications, but an alternative expression in terms of metric tensor components can be derived from them and is very useful. Since the metric tensor $g_{\mu\nu}$ is a rank-2 covariant tensor (as shown in Chapter 4), it is related to the Minkowski metric $\eta_{\alpha\beta}$ of flat space-time by

$$g_{\mu\nu} = \eta_{\alpha\beta}\frac{\partial \xi^\alpha}{\partial x^\mu}\frac{\partial \xi^\beta}{\partial x^\nu} . \tag{A.3}$$

Now differentiate this with respect to x^λ :

$$\frac{\partial g_{\mu\nu}}{\partial x^\lambda} = \frac{\partial^2 \xi^\alpha}{\partial x^\mu \partial x^\lambda} \frac{\partial \xi^\beta}{\partial x^\nu} \eta_{\alpha\beta} + \frac{\partial \xi^\alpha}{\partial x^\mu} \frac{\partial^2 \xi^\beta}{\partial x^\nu \partial x^\lambda} \eta_{\alpha\beta} \tag{A.4}$$

(since $\eta_{\alpha\beta}$ is a constant). From Equation (A.2):

$$\Gamma^\rho_{\mu\nu} \frac{\partial \xi^\beta}{\partial x^\rho} = \overbrace{\frac{\partial x^\rho}{\partial \xi^\alpha} \frac{\partial \xi^\beta}{\partial x^\rho}}^{\delta^\beta_\alpha} \frac{\partial^2 \xi^\alpha}{\partial x^\nu \partial x^\mu} = \frac{\partial^2 \xi^\beta}{\partial x^\nu \partial x^\mu} .$$

Using this to eliminate the second derivatives in Equation (A.4),

$$\frac{\partial g_{\mu\nu}}{\partial x^\lambda} = \Gamma^\rho_{\mu\lambda} \underbrace{\frac{\partial \xi^\alpha}{\partial x^\rho} \frac{\partial \xi^\beta}{\partial x^\nu} \eta_{\alpha\beta}}_{g_{\rho\nu}} + \Gamma^\rho_{\nu\lambda} \underbrace{\frac{\partial \xi^\beta}{\partial x^\rho} \frac{\partial \xi^\alpha}{\partial x^\mu} \eta_{\alpha\beta}}_{g_{\rho\mu}},$$

where the underbraced expressions are just the metric tensor as given by Equation (A.3). Thus,

$$\frac{\partial g_{\mu\nu}}{\partial x^\lambda} = \Gamma^\rho_{\mu\lambda} g_{\rho\nu} + \Gamma^\rho_{\nu\lambda} g_{\rho\mu} . \tag{A.5}$$

We now write this expression three times: once as is, once with μ and λ interchanged, and once with ν and λ interchanged:

$$\begin{aligned}
\frac{\partial g_{\mu\nu}}{\partial x^\lambda} &= \overbrace{\Gamma^\rho_{\mu\lambda} g_{\rho\nu}}^{I} + \overbrace{\Gamma^\rho_{\nu\lambda} g_{\rho\mu}}^{III} , \\
\frac{\partial g_{\lambda\nu}}{\partial x^\mu} &= \overbrace{\Gamma^\rho_{\mu\lambda} g_{\rho\nu}}^{I} + \overbrace{\Gamma^\rho_{\nu\mu} g_{\rho\lambda}}^{II} , \\
\frac{\partial g_{\mu\lambda}}{\partial x^\nu} &= \overbrace{\Gamma^\rho_{\mu\nu} g_{\rho\lambda}}^{II} + \overbrace{\Gamma^\rho_{\nu\lambda} g_{\rho\mu}}^{III} ,
\end{aligned}$$

where we have made use of the symmetry of Γ and g in their lower indices, and have identified three pairs of identical expressions with overbraces. Add the first two of these equations and subtract the third to obtain

$$2\Gamma^\rho_{\mu\lambda} g_{\rho\nu} = \frac{\partial g_{\mu\nu}}{\partial x^\lambda} + \frac{\partial g_{\lambda\nu}}{\partial x^\mu} - \frac{\partial g_{\mu\lambda}}{\partial x^\nu} .$$

We now multiply both sides of the equation by the contravariant form $g^{\nu\sigma}$ of the metric tensor, which has the matrix inverse property

$$g_{\rho\nu} g^{\nu\sigma} = \delta^\sigma_\rho ,$$

so

$$\Gamma^{\sigma}_{\mu\lambda} = \frac{1}{2} g^{\nu\sigma} \left[\frac{\partial g_{\mu\nu}}{\partial x^{\lambda}} + \frac{\partial g_{\lambda\nu}}{\partial x^{\mu}} - \frac{\partial g_{\mu\lambda}}{\partial x^{\nu}} \right]. \tag{A.6}$$

This metric expression is usually the most convenient form of the affine connection.

Transformations

Note from Equation (A.6) that in an LIRF with constant metric tensor components (e.g., Cartesian), Γ will be identically zero and, from Equation (5.9), $d^2x^{\lambda}/d\tau^2 = 0$ in this LIRF. If Γ were a tensor, the Principle of Covariance would then imply that this same equation applied in *all* coordinate systems so that no acceleration would occur in curved space-time, or in non-inertial reference frames. This is clearly nonsense, and shows that the affine connection cannot be a tensor. The form of its transformation under changes of coordinate systems is more complicated than that of a tensor.

To see how the affine connection transforms, begin with the coordinate form of the affine connection, Equation (A.2), and remember that, e.g.,

$$\frac{\partial V}{\partial x^{\mu}} = \frac{\partial V}{\partial \bar{x}^{\nu}} \frac{\partial \bar{x}^{\nu}}{\partial x^{\mu}}, \tag{A.7}$$

for some other coordinate system $\{\bar{x}\}$. Thus, Γ transforms to $\bar{\Gamma}$ under the change of variables $\{x\} \to \{\bar{x}\}$ as

$$\begin{aligned}
\bar{\Gamma}^{\lambda}_{\mu\nu} &= \frac{\partial \bar{x}^{\lambda}}{\partial \xi^{\alpha}} \frac{\partial^2 \xi^{\alpha}}{\partial \bar{x}^{\mu} \partial \bar{x}^{\nu}}, \\
&= \frac{\partial \bar{x}^{\lambda}}{\partial x^{\rho}} \frac{\partial x^{\rho}}{\partial \xi^{\alpha}} \frac{\partial}{\partial \bar{x}^{\mu}} \left(\frac{\partial x^{\sigma}}{\partial \bar{x}^{\nu}} \frac{\partial \xi^{\alpha}}{\partial x^{\sigma}} \right), \\
&= \frac{\partial \bar{x}^{\lambda}}{\partial x^{\rho}} \frac{\partial x^{\rho}}{\partial \xi^{\alpha}} \left[\frac{\partial x^{\sigma}}{\partial \bar{x}^{\nu}} \frac{\partial^2 \xi^{\alpha}}{\partial \bar{x}^{\mu} \partial x^{\sigma}} + \frac{\partial^2 x^{\sigma}}{\partial \bar{x}^{\mu} \partial \bar{x}^{\nu}} \frac{\partial \xi^{\alpha}}{\partial x^{\sigma}} \right], \\
&= \frac{\partial \bar{x}^{\lambda}}{\partial x^{\rho}} \frac{\partial x^{\rho}}{\partial \xi^{\alpha}} \left[\frac{\partial x^{\sigma}}{\partial \bar{x}^{\nu}} \frac{\partial x^{\tau}}{\partial \bar{x}^{\mu}} \frac{\partial^2 \xi^{\alpha}}{\partial x^{\tau} \partial x^{\sigma}} + \frac{\partial^2 x^{\sigma}}{\partial \bar{x}^{\mu} \partial \bar{x}^{\nu}} \frac{\partial \xi^{\alpha}}{\partial x^{\sigma}} \right], \\
&= \frac{\partial \bar{x}^{\lambda}}{\partial x^{\rho}} \frac{\partial x^{\tau}}{\partial \bar{x}^{\mu}} \frac{\partial x^{\sigma}}{\partial \bar{x}^{\nu}} \left[\frac{\partial x^{\rho}}{\partial \xi^{\alpha}} \frac{\partial^2 \xi^{\alpha}}{\partial x^{\tau} \partial x^{\sigma}} \right] + \frac{\partial \bar{x}^{\lambda}}{\partial x^{\rho}} \underbrace{\frac{\partial x^{\rho}}{\partial \xi^{\alpha}} \frac{\partial \xi^{\alpha}}{\partial x^{\sigma}}}_{\delta^{\rho}_{\sigma}} \frac{\partial^2 x^{\sigma}}{\partial \bar{x}^{\mu} \partial \bar{x}^{\nu}},
\end{aligned}$$

where we have used Equation (A.7) several times. The quantity in brackets in the last line is the same as that in the coordinate form of the affine connection (Equation (A.2), q.v.), and the delta function in the second term changes σ to ρ, so we see that

$$\bar{\Gamma}^{\lambda}_{\mu\nu} = \frac{\partial \bar{x}^{\lambda}}{\partial x^{\rho}} \frac{\partial x^{\kappa}}{\partial \bar{x}^{\mu}} \frac{\partial x^{\sigma}}{\partial \bar{x}^{\nu}} \Gamma^{\rho}_{\kappa\sigma} + \frac{\partial \bar{x}^{\lambda}}{\partial x^{\rho}} \frac{\partial^2 x^{\rho}}{\partial \bar{x}^{\mu}\, \partial \bar{x}^{\nu}} . \tag{A.8}$$

The first term on the right is exactly what would be expected if Γ were a mixed, rank-3 tensor; the second term ruins it. The affine connection does not transform as a tensor.

There is an alternative form of this transformation relation that is often useful. Start with the identity

$$\frac{\partial \bar{x}^{\lambda}}{\partial x^{\rho}} \frac{\partial x^{\rho}}{\partial \bar{x}^{\nu}} = \delta^{\lambda}_{\nu}$$

and differentiate with respect to $\bar{x}^{\mu}$. The right-hand side is a constant with 0 derivative, so the result is just

$$\frac{\partial \bar{x}^{\lambda}}{\partial x^{\rho}} \frac{\partial^2 x^{\rho}}{\partial \bar{x}^{\mu}\, \partial \bar{x}^{\nu}} = -\frac{\partial x^{\rho}}{\partial \bar{x}^{\nu}} \frac{\partial^2 x^{\lambda}}{\partial \bar{x}^{\mu} \partial x^{\rho}} .$$

Comparing this to Equation (A.8), we have the alternative transformation relation

$$\bar{\Gamma}^{\lambda}_{\mu\nu} = \frac{\partial \bar{x}^{\lambda}}{\partial x^{\rho}} \frac{\partial x^{\kappa}}{\partial \bar{x}^{\mu}} \frac{\partial x^{\sigma}}{\partial \bar{x}^{\nu}} \Gamma^{\rho}_{\kappa\sigma} - \frac{\partial x^{\rho}}{\partial \bar{x}^{\nu}} \frac{\partial^2 x^{\lambda}}{\partial x^{\rho}\, \partial \bar{x}^{\mu}} . \tag{A.9}$$

A.2 Geodesic paths

The equations of motion of a freely falling particle can be shown to be the shortest path between its endpoints in curved space-time; i.e., a *geodesic*. In flat space this is a straight line; on the curved two-dimensional surface of a 3-sphere it is a great circle; and, in general, it is not a straight line if the geometry is not Euclidean. Now, the four-dimensional space-time distance between points is the line element or proper length:

$$ds^2 = g_{\mu\nu} dx^{\mu} dx^{\nu},$$

where $dx^0 = cdt$. We can equally write this distance as the *proper time*:

$$d\tau^2 = -g_{\mu\nu} dx^{\mu} dx^{\nu} ,$$

so $d\tau^2 > 0$ for speeds less than that of light, and $= 0$ for light. We formulate a geodesic path as one of least proper time, and seek to minimize

$$\tau_{BA} = \int_A^B d\tau \ ,$$

where A and B are the fixed end points of the path. Let p be any parameter along a time-like path through space-time (e.g., arc length), so

$$d\tau = \frac{d\tau}{dp}dp = \frac{\sqrt{-g_{\mu\nu}dx^\mu dx^\nu}}{dp}dp = \sqrt{-g_{\mu\nu}\frac{dx^\mu}{dp}\frac{dx^\nu}{dp}}dp \ , \tag{A.10}$$

and we seek a path $x^\lambda\,(p)$ that minimizes

$$\tau_{BA} = \int_A^B \sqrt{-g_{\mu\nu}\frac{dx^\mu}{dp}\frac{dx^\nu}{dp}}dp \ .$$

The solution comes from the calculus of variations. Let the minimum-length path $x^\lambda\,(p)$ be perturbed infinitesimally to $x^\lambda\,(p) + \delta x^\lambda\,(p)$; analogous to ordinary calculus we require that τ_{BA} not be changed by this perturbation, if it is truly the minimal path. The change in τ_{BA} will be of the form

$$\begin{aligned}\delta\tau &= \int \delta\sqrt{-f}dp \ , \\ &= -\frac{1}{2}\int (-f)^{-1/2}\,\delta f\,dp \ ,\end{aligned}$$

with $f = g_{\mu\nu}\,(dx^\mu/dp)\,(dx^\nu/dp)$. If we now set the arbitrary parameter $p = \tau$, we have from Equation (A.10) that $(-f)^{-1/2} = -d\tau/d\tau = -1$; so the variational integral may be written as

$$\delta\tau_{BA} = \frac{1}{2}\int_A^B \delta f\,d\tau = \frac{1}{2}\int_A^B \delta\left(g_{\mu\nu}\frac{dx^\mu}{d\tau}\frac{dx^\nu}{d\tau}\right)d\tau \ .$$

This can be evaluated exactly but, since we are concerned only with infinitesimal changes in the path, it is legitimate to approximate things accordingly; and this greatly simplifies the calculation. Thus, we set $\delta g_{\mu\nu} \to (\partial g_{\upsilon\nu}/\partial x^\kappa)\,\delta x^\kappa$ (note the implied sum!) and the variational integral becomes

$$\begin{aligned}\delta\tau_{BA} = \frac{1}{2}\int_A^B &\left[\frac{\partial g_{\mu\nu}}{\partial x^\kappa}\frac{dx^\mu}{d\tau}\frac{dx^\nu}{d\tau}\delta x^\kappa\right. \\ &\left.+g_{\mu\nu}\frac{d\,(\delta x^\mu)}{d\tau}\frac{dx^\nu}{d\tau} + g_{\mu\nu}\frac{dx^\mu}{d\tau}\frac{d\,(\delta x^\nu)}{d\tau}\right]d\tau \ .\end{aligned} \tag{A.11}$$

We can integrate the second and third terms by parts; e.g.,

$$\frac{1}{2}\int_A^B g_{\mu\nu}\frac{dx^\mu}{d\tau}\frac{d\,(\delta x^\nu)}{d\tau} = \frac{1}{2}\int_A^B U\,dV,$$

where

$$U = g_{\mu\nu}\frac{dx^{\mu}}{d\tau} \Rightarrow dU = \left(\frac{dg_{\mu\nu}}{d\tau}\frac{dx^{\mu}}{d\tau} + g_{\mu\nu}\frac{d^2x^{\mu}}{d\tau^2}\right)d\tau \ ,$$

$$dV = \frac{d\left(\delta x^{\nu}\right)}{d\tau}d\tau \Rightarrow V = \delta x^{\nu} \ ,$$

so that

$$\int_A^B U\,dV = g_{\mu\nu}\frac{dx^{\mu}}{d\tau}\delta x^{\upsilon}\Bigg]_A^B - \int_A^B\left[\frac{dg_{\mu\nu}}{d\tau}\frac{dx^{\mu}}{d\tau} + g_{\mu\nu}\frac{d^2x^{\mu}}{d\tau^2}\right]\delta x^{\nu}\,d\tau \ .$$

The first term on the right-hand side is zero since δx vanishes at the two endpoints, and for the first term inside the integral we can once again approximate the derivative of g as $dg_{\mu\nu} = \left(\partial g_{\mu\nu}/\partial x^{\kappa}\right)\delta x^{\kappa}$, so the second term in Equation (A.11) is

$$\frac{1}{2}\int_A^B g_{\mu\nu}\frac{dx^{\mu}}{d\tau}\frac{d\left(\delta x^{\nu}\right)}{d\tau} = -\frac{1}{2}\int_A^B\left[\frac{\partial g_{\mu\nu}}{\partial x^{\kappa}}\frac{dx^{\mu}}{d\tau}\frac{dx^{\kappa}}{d\tau} + g_{\mu\nu}\frac{d^2x^{\mu}}{d\tau^2}\right]\delta x^{\nu}\,d\tau \ .$$

We can pull a similar trick on the last term of Equation (A.11) so that the variational integral becomes (after some artful rearrangement of indices)

$$\delta\tau_{BA} = -\int_A^B\left[g_{\mu\kappa}\frac{d^2x^{\mu}}{d\tau^2} + \frac{1}{2}\left(\frac{\partial g_{\nu\kappa}}{\partial x^{\mu}} + \frac{\partial g_{\mu\kappa}}{\partial x^{\nu}} - \frac{\partial g_{\mu\nu}}{\partial x^{\kappa}}\right)\frac{dx^{\mu}}{d\tau}\frac{dx^{\nu}}{d\tau}\right]\delta x^{\kappa}\,d\tau \ .$$

If this is to be zero for all infinitesimal displacements δx^{κ}, the quantity in braces must be identically zero:

$$g_{\mu\kappa}\frac{d^2x^{\mu}}{d\tau^2} + \frac{1}{2}\left(\frac{\partial g_{\nu\kappa}}{\partial x^{\mu}} + \frac{\partial g_{\mu\kappa}}{\partial x^{\nu}} - \frac{\partial g_{\mu\nu}}{\partial x^{\kappa}}\right)\frac{dx^{\mu}}{d\tau}\frac{dx^{\nu}}{d\tau} = 0$$

for all κ. Multiplying by the inverse of the metric tensor $g^{\lambda\kappa}$ and summing over κ, the first term becomes

$$g^{\lambda\kappa}g_{\mu\kappa}\frac{d^2x^{\mu}}{d\tau^2} = \delta^{\lambda}_{\mu}\frac{d^2x^{\mu}}{d\tau^2} = \frac{d^2x^{\lambda}}{d\tau^2}$$

and the entire equation is

$$\frac{d^2x^{\lambda}}{d\tau^2} + \frac{1}{2}g^{\lambda\kappa}\left[\frac{\partial g_{\nu\kappa}}{\partial x^{\mu}} + \frac{\partial g_{\mu\kappa}}{\partial x^{\nu}} - \frac{\partial g_{\mu\nu}}{\partial x^{\kappa}}\right]\frac{dx^{\mu}}{d\tau}\frac{dx^{\nu}}{d\tau} = 0 \ .$$

But from the definition of the affine connection (Equation (A.6)),

$$\Gamma^{\lambda}_{\mu\nu} = \frac{1}{2}g^{\lambda\kappa}\left[\frac{\partial g_{\nu\kappa}}{\partial x^{\mu}} + \frac{\partial g_{\mu\kappa}}{\partial x^{\nu}} - \frac{\partial g_{\mu\nu}}{\partial x^{\kappa}}\right],$$

so the variational solution is just the Equations of Motion of a freely falling particle:

$$\frac{d^2x^\lambda}{d\tau^2} + \Gamma^\lambda_{\mu\kappa}\frac{dx^\mu}{d\tau}\frac{dx^\nu}{d\tau} = 0 \,. \tag{A.12}$$

Thus: *a freely falling particle follows a geodesic trajectory in curved space-time*, given by the above geodesic equation. We note that there is one complication that we have glossed over: for light $d\tau = 0$ so this equation does not apply. Instead, we say that light follows a *null* geodesic, one of zero path length as measured by proper time:

$$d\tau^2 = -ds^2 = 0 \qquad \text{(light)} \,.$$

A.3 Covariant differentiation

Coordinate differentiation is not generally covariant

A contravariant vector transforms as, e.g.,

$$\bar{V}^\mu = \frac{\partial \bar{x}^\mu}{\partial x^\nu} V^\nu \,. \tag{A.13}$$

Differentiating both sides with respect to $\bar{x}^\lambda$ gives

$$\begin{aligned}
\frac{\partial \bar{V}^\mu}{\partial \bar{x}^\lambda} &= \frac{\partial}{\partial \bar{x}^\lambda}\left(\frac{\partial \bar{x}^\mu}{\partial x^\nu} V^\nu\right) \\
&= \frac{\partial \bar{x}^\mu}{\partial x^\nu}\frac{\partial V^\nu}{\partial \bar{x}^\lambda} + \frac{\partial^2 \bar{x}^\mu}{\partial x^\nu \partial \bar{x}^\lambda} V^\nu \\
&= \frac{\partial \bar{x}^\mu}{\partial x^\nu}\frac{\partial x^\rho}{\partial \bar{x}^\lambda}\frac{\partial V^\nu}{\partial x^\rho} + \frac{\partial^2 \bar{x}^\mu}{\partial x^\nu \partial \bar{x}^\lambda} V^\nu \,,
\end{aligned}$$

where we have used the chain rule

$$\frac{\partial}{\partial \bar{x}^\lambda} = \frac{\partial x^\rho}{\partial \bar{x}^\lambda}\frac{\partial}{\partial x^\rho}$$

in the last step in order to rationalize the first term on the right. Thus, the rule for transforming coordinate derivatives of contravariant vectors is

$$\frac{\partial \bar{V}^\mu}{\partial \bar{x}^\lambda} = \frac{\partial \bar{x}^\mu}{\partial x^\nu}\frac{\partial x^\rho}{\partial \bar{x}^\lambda}\frac{\partial V^\nu}{\partial x^\rho} + \overbrace{\frac{\partial^2 \bar{x}^\mu}{\partial x^\rho \partial \bar{x}^\lambda} V^\rho} \,. \tag{A.14}$$

The first term on the right is the expected transformation for a mixed rank-2 tensor, but the second term (with overbrace) is extraneous to a tensor transformation and ruins the supposed tensor nature of a coordinate derivative.

Covariant derivatives of a contravariant tensor

The problem arises from the fact that coordinate differentiation involves changes not only of tensor components, but of the underlying coordinate bases (including effects of curvature). To see how component differentiation incorporates basis changes we employ the basis representation for vector components: $V^\mu = \left(\hat{\mathbf{e}} \cdot \vec{\mathbf{V}}\right)^\mu$, so

$$\frac{\partial}{\partial x^\nu} V^\mu = \frac{\partial}{\partial x^\nu}\left(\hat{\mathbf{e}} \cdot \vec{\mathbf{V}}\right)^\mu = \hat{\mathbf{e}}^\mu \cdot \frac{\partial \vec{\mathbf{V}}}{\partial x^\nu} + \vec{\mathbf{V}} \cdot \frac{\partial \hat{\mathbf{e}}^\mu}{\partial x^\nu} .$$

Since the last term on the right encapsulates the change in basis vectors, we tentatively take as the **covariant derivative** with respect to x^ν, denoted D_ν, the ordinary derivative minus this term:[1]

$$D_\nu V^\mu \to \frac{\partial V^\mu}{\partial x^\nu} - \vec{\mathbf{V}} \cdot \frac{\partial \hat{\mathbf{e}}^\mu}{\partial x^\nu} . \tag{A.15}$$

This will, hopefully, get rid of the non-tensor second term in Equation (A.14). From Equation (A.1) for the affine connection we have

$$\frac{\partial \hat{\mathbf{e}}_\mu}{\partial x^\nu} = \Gamma^\lambda_{\mu\nu} \hat{\mathbf{e}}_\lambda . \tag{A.16}$$

Employing the identity $\hat{\mathbf{e}}_\lambda \cdot \hat{\mathbf{e}}^\lambda = 1$ and the symmetry of Γ in its lower indices, this affine connection relation may be written in terms of contravariant bases as

$$\frac{\partial \hat{\mathbf{e}}^\mu}{\partial x^\nu} = -\Gamma^\mu_{\nu\lambda} \hat{\mathbf{e}}^\lambda . \tag{A.17}$$

Incorporating this expression into that for the covariant derivative,

$$D_\nu V^\mu = \frac{\partial V^\mu}{\partial x^\nu} + \Gamma^\mu_{\nu\lambda} V^\lambda . \tag{A.18}$$

Note that this expression reduces to ordinary coordinate differentiation in flat geometries where coordinate systems may be found in which the affine connection components are identically zero; this is a necessary condition of covariant differentiation.

It remains to show that this quantity is generally covariant. The transformation property of the (non-tensor) affine connection is, from Equation (A.9),

$$\bar{\Gamma}^\mu_{\lambda\kappa} = \frac{\partial \bar{x}^\mu}{\partial x^\rho} \frac{\partial x^\tau}{\partial \bar{x}^\lambda} \frac{\partial x^\sigma}{\partial \bar{x}^\kappa} \Gamma^\rho_{\tau\sigma} - \frac{\partial x^\rho}{\partial \bar{x}^\kappa} \frac{\partial^2 \bar{x}^\mu}{\partial x^\rho \partial \bar{x}^\lambda} .$$

[1] Denoted here by the symbol D; more commonly denoted in the literature by a covariant semi-colon; e.g., $V_{\mu;\lambda} \equiv D_\lambda V_\mu$.

Multiplying this by the transformation of the contravariant vector V (Equation (A.13)),

$$\bar{\Gamma}^{\mu}_{\lambda\kappa}\bar{V}^{\kappa} = \frac{\partial \bar{x}^{\mu}}{\partial x^{\nu}}\frac{\partial x^{\rho}}{\partial \bar{x}^{\lambda}}\Gamma^{\nu}_{\rho\sigma}V^{\sigma} - \overbrace{\frac{\partial^2 \bar{x}^{\mu}}{\partial \bar{x}^{\lambda}\partial x^{\sigma}}V^{\sigma}}\,, \tag{A.19}$$

where we have twice used the identity

$$\frac{\partial x^{\sigma}}{\partial \bar{x}^{\kappa}}\frac{\partial \bar{x}^{\kappa}}{\partial x^{\eta}}V^{\eta} = \delta^{\sigma}_{\eta}V^{\eta} = V^{\sigma}\,.$$

If we now add Equations (A.14) and (A.19), the inhomogeneous terms (those with overbraces) cancel and we have

$$\frac{\partial \bar{V}^{\mu}}{\partial \bar{x}^{\lambda}} + \bar{\Gamma}^{\mu}_{\lambda\kappa}\bar{V}^{\kappa} = \frac{\partial \bar{x}^{\mu}}{\partial x^{\nu}}\frac{\partial x^{\rho}}{\partial \bar{x}^{\lambda}}\left[\frac{\partial V^{\nu}}{\partial x^{\rho}} + \Gamma^{\nu}_{\rho\sigma}V^{\sigma}\right]\,.$$

This is the transformation rule for a mixed, rank-2 tensor; and it agrees with the ordinary coordinate derivative in an LIRF, or flat space-time, where Γ is zero. We thus adopt this as our definition of covariant derivative, and the *covariant derivative of a contravariant vector* is

$$D_{\lambda}V^{\mu} = \frac{\partial V^{\mu}}{\partial x^{\lambda}} + \Gamma^{\mu}_{\lambda\kappa}V^{\kappa}\,. \tag{A.20}$$

Covariant derivatives of other tensor forms

Similarly, the *covariant derivative of a covariant vector* can be shown to be given by

$$D_{\lambda}V_{\mu} = \frac{\partial V_{\mu}}{\partial x^{\lambda}} - \Gamma^{\kappa}_{\lambda\mu}V_{\kappa}\,. \tag{A.21}$$

Compare the last two equations: in both cases, covariant differentiation adds one covariant index to the vector, and the only difference between derivatives of covariant and contravariant vectors is in the sign of the affine connection terms (with summation indices as appropriate). With this in mind, the covariant derivative of a higher rank tensor is just (1) the ordinary partial derivative, plus (2) a $+\Gamma$ term for each contravariant index (as in Equation (A.20)), and (3) a $-\Gamma$ term for each covariant index (as in Equation (A.21)). Here are examples:

$$D_{\nu}T^{\alpha\beta} = \frac{\partial T^{\alpha\beta}}{\partial x^{\nu}} + \Gamma^{\alpha}_{\nu\lambda}T^{\lambda\beta} + \Gamma^{\beta}_{\nu\lambda}T^{\alpha\lambda}\,, \tag{A.22}$$

$$D_\nu g_{\alpha\beta} = \frac{\partial g_{\alpha\beta}}{\partial x^\nu} - \Gamma^\lambda_{\alpha\nu} g_{\lambda\beta} - \Gamma^\lambda_{\nu\beta} g_{\alpha\lambda} , \tag{A.23}$$

$$D_\nu A^\mu_\rho = \frac{\partial A^\mu_\rho}{\partial x^\nu} + \Gamma^\lambda_{\nu\rho} A^\mu_\lambda - \Gamma^\mu_{\lambda\nu} A^\lambda_\rho . \tag{A.24}$$

If you study these carefully you will see that there is only one consistent way of assigning indices to the Γs, so you can usually figure out the form for any particular tensor. Note that, in every case, the resulting derivative is a tensor of one covariant rank higher than the original tensor; e.g., $D_\nu g_{\alpha\beta}$ is a rank-3 covariant tensor, and $D_\nu V^\mu$ is a rank-2 mixed tensor.

Covariant divergence

The divergence of a vector in 'ordinary' calculus is just $\operatorname{div} \vec{V} \equiv \vec{\nabla} \cdot \vec{V} = \sum_i \partial V^i / \partial x^i$ (in Cartesian coordinates). The obvious generalization to contravariant tensors of any rank is

$$\operatorname{div} T^{\cdots} = D_\mu T^{\cdot\cdot\mu} . \tag{A.25}$$

Of particular importance in GR are contravariant divergences of rank-2 contravariant tensors:

$$(\operatorname{div} T)^{\cdot\nu} = D_\mu T^{\mu\nu} . \tag{A.26}$$

From Equation (A.25) the divergence of a rank-1 tensor ('vector ') is a scalar. One can show that its value is given by the simple expression

$$D_\mu V^\mu = \frac{1}{\sqrt{-g}} \frac{\partial}{\partial x^\mu} \left(\sqrt{-g} V^\mu \right) ,$$

where $g \equiv \det(g_{\lambda\kappa})$ is the determinant of the metric tensor matrix.

Finally, note that the covariant derivative of the metric tensor is identically zero, as can be seen from

$$D_\lambda g_{\mu\nu} = \frac{\partial g_{\mu\nu}}{\partial x^\lambda} - \Gamma^\kappa_{\lambda\mu} g_{\kappa\nu} - \Gamma^\kappa_{\nu\lambda} g_{\kappa\mu} .$$

But we previously (Equation (A.5)) showed that

$$\frac{\partial g_{\mu\nu}}{\partial x^\lambda} = \Gamma^\kappa_{\lambda\mu} g_{\kappa\nu} + \Gamma^\kappa_{\nu\lambda} g_{\kappa\mu}$$

so that $D_\lambda g_{\mu\nu} = 0$ identically.

A.4 Curvature tensors

The Gaussian curvature formula of Chapter 6 is only applicable to surfaces of two dimensions or fewer, and only of constant curvature. The more general connection between curvature and the metric tensor components was developed by his student, George Friedrich Riemann. It is this very general form that is required by GR.

Curvature revealed

Defining vectors in terms of unit vectors as in the previous section, the change in the components of $\vec{V}$ when it moves between two infinitesimally nearby points is

$$\Delta V^{\mu} = \left[\Delta\vec{V}\cdot\hat{e}^{\mu}\right] + \left[\vec{V}\cdot\Delta\hat{e}^{\mu}\right]. \tag{A.27}$$

The first of these bracketed expressions arises from the change in $\vec{V}$ itself, the second from the change in coordinate bases over the path taken by $\vec{V}$. It is this second term that reflects the curvature of the underlying geometry. The effect of curvature reveals itself most clearly in the phenomenon of *parallel transport*.

Consider the surface of a sphere in 3-space. Start at the north pole with a vector pointing straight down the meridian, and move the vector down to the equator without rotating it. Now move it along the equator by 90 degrees, again keeping it always parallel to itself (pointing straight south). Finally, move it back up its new meridian to the pole, and observe that it will have been rotated 90 degrees with respect to its original orientation, even though it was moved parallel to itself all along its path (i.e., its orientation after moving an infinitesimal distance was always parallel to its starting position). This rotation upon parallel transport occurs if, and only if, the surface is curved. Riemann thus took as a measure of curvature of any space, the amount of vector rotation upon parallel transport about an infinitesimal closed path. This will be the integral of the second bracketed expression in Equation A.27 . That term – the change in V^{μ} due only to changes in the coordinate system – will be expected to be proportional to the original vector components and to the displacement. From Equation (A.17) for the affine connection we surmise that

$$\Delta V^{\mu}_{\parallel} = -\Gamma^{\mu}_{\nu\kappa} V^{\kappa} \Delta x^{\nu} \tag{A.28}$$

(in the limit as $\Delta x \to 0$), where the $\parallel$ subscript denotes parallel transport. This is the change in V^{μ} due to parallel transport along the infinitesimal path given by the Δx^{ν}.

Riemann Curvature Tensor

Here is a sketch of the derivation of the Riemann Curvature Tensor as a measure of vector rotation upon parallel transport around a closed path. We set up the path to consist of motion along two coordinate axes (see Figure A.1). Then for translation

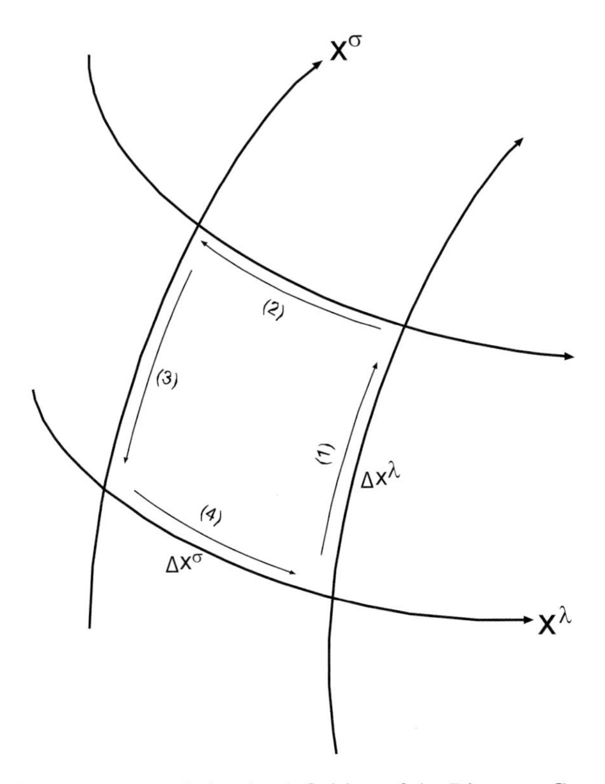

Figure A.1 Parallel transport path for the definition of the Riemann Curvature Tensor.

along one of these axes (x^σ) we have (due only to the coordinate system)

$$\frac{\partial V^\mu_{\shortparallel}}{\partial x^\lambda} = -\Gamma^\mu_{\lambda\kappa} V^\kappa \,, \tag{A.29}$$

$$\Rightarrow V^\mu_{\shortparallel} \text{ (final)} = V^\mu \text{ (original)} - \int_{x_0^\lambda}^{x_0^\lambda+\delta x^\lambda} \Gamma^\mu_{\lambda\kappa} V^\kappa dx^\lambda$$

(no sum on λ); this change corresponds to parallel transport ($\|$) along path (1) in the figure. We now translate along the other axis x^λ (path (2) in the figure), then back along x^σ at the new value of x^λ (path (3)); then back along the x^λ axis starting with the last value of x^σ (path (4)). Subtracting the changes along the x^σ axis at two values of the other coordinate – i.e., along opposite sides of the 'rectangular' loop – we get

$$\Delta V^\mu_{\shortparallel} = -\left[\int_{x_0^\lambda}^{x_0^\lambda+\delta x^\lambda} \Gamma^\mu_{\lambda\kappa} V^\kappa dx^\lambda\right]_{x^\sigma = x_0^\sigma} + \left[\int_{x_0^\lambda+\delta x^\lambda}^{x_0^\lambda} \Gamma^\mu_{\lambda\kappa} V^\kappa dx^\lambda\right]_{x^\sigma = x_0^\sigma+\delta x^\sigma}$$

$$\approx -\delta x^\sigma \int_{x_0^\lambda}^{x_0^\lambda+\delta x^\lambda} \frac{\partial}{\partial x^\sigma}\left(\Gamma^\mu_{\lambda\kappa} V^\kappa\right) dx^\lambda$$

(no sums on λ, σ). Doing the same for the other two sides of the loop, we have the total change in V^μ from parallel transport as

$$\Delta V^\mu_{\parallel} = -\delta x^\sigma \int \frac{\partial}{\partial x^\sigma}\left(\Gamma^\mu_{\lambda\kappa} V^\kappa\right) dx^\lambda + \delta x^\lambda \int \frac{\partial}{\partial x^\lambda}\left(\Gamma^\mu_{\lambda\kappa} V^\kappa\right) dx^\sigma$$
$$\approx \delta x^\sigma\, \delta x^\lambda \left[-\frac{\partial}{\partial x^\sigma}\left(\Gamma^\mu_{\lambda\kappa} V^\kappa\right) + \frac{\partial}{\partial x^\lambda}\left(\Gamma^\mu_{\lambda\kappa} V^\kappa\right)\right] \tag{A.30}$$

(no sums on λ, σ), where we have approximated the integrals by products of integrands and the (infinitesimal) lengths of integration paths, δx. This last expression is thus the change in vector component V^μ due entirely to parallel transport around an infinitesimally small path encompassed by δx^λ and δx^σ.

We carry through the derivatives in this expression, using Equation (A.29) to evaluate the derivatives of V^μ; the result is

$$\Delta V^\mu_{\parallel} = \delta x^\sigma \delta x^\lambda \left[\frac{\partial \Gamma^\mu_{\kappa\sigma}}{\partial x^\lambda} - \frac{\Gamma^\mu_{\kappa\lambda}}{\partial x^\sigma} + \Gamma^\mu_{\nu\lambda}\Gamma^\nu_{\kappa\sigma} - \Gamma^\mu_{\nu\sigma}\Gamma^\nu_{\kappa\lambda}\right] V^\kappa$$

(no sums on σ, λ). Not apparent from this derivation is that the quantity in brackets is, indeed, a tensor (as it must be since it reflects a quantity – vector rotation – independent of the coordinate system). Since vector rotation around a closed path is proportional to this quantity, and that rotation occurs if and only if space-time is curved, we use this quantity to *define* curvature of space-time. This is the **Riemann Curvature Tensor**:

$$\mathcal{R}^\lambda_{\ \mu\nu\kappa} \equiv \frac{\partial \Gamma^\lambda_{\mu\nu}}{\partial x^\kappa} - \frac{\partial \Gamma^\lambda_{\mu\kappa}}{\partial x^\nu} + \Gamma^\eta_{\mu\nu}\Gamma^\lambda_{\kappa\eta} - \Gamma^\eta_{\mu\kappa}\Gamma^\lambda_{\nu\eta}\,. \tag{A.31}$$

The Riemann Curvature Tensor is the simplest tensor capable of fully characterizing curvature in four-dimensional geometries. This tensor is often used in its fully covariant form:

$$\mathcal{R}_{\lambda\mu\nu\kappa} = g_{\lambda\sigma}\mathcal{R}^\sigma_{\ \mu\nu\kappa}\,,$$
$$= g_{\lambda\sigma}\left[\frac{\partial \Gamma^\sigma_{\mu\nu}}{\partial x^\kappa} - \frac{\partial \Gamma^\sigma_{\mu\kappa}}{\partial x^\nu} + \Gamma^\eta_{\mu\nu}\Gamma^\sigma_{\kappa\eta} - \Gamma^\eta_{\mu\kappa}\Gamma^\sigma_{\nu\eta}\right]. \tag{A.32}$$

One can show (after a tedious calculation) that an alternative representation for this tensor is

$$\mathcal{R}_{\lambda\mu\nu\kappa} = \frac{1}{2}\left[\frac{\partial^2 g_{\lambda\nu}}{\partial x^\mu\,\partial x^\kappa} - \frac{\partial^2 g_{\mu\nu}}{\partial x^\lambda\,\partial x^\kappa} - \frac{\partial^2 g_{\lambda\kappa}}{\partial x^\mu\,\partial x^\nu} + \frac{\partial^2 g_{\mu\kappa}}{\partial x^\lambda\,\partial x^\nu}\right]$$
$$+ g_{\eta\sigma}\left[\Gamma^\eta_{\nu\lambda}\Gamma^\sigma_{\mu\kappa} - \Gamma^\eta_{\kappa\lambda}\Gamma^\sigma_{\mu\nu}\right]. \tag{A.33}$$

From either of the above two equations, the following symmetries are apparent:

Symmetry of index pairs

$$\mathcal{R}_{\lambda\mu\nu\kappa} = \mathcal{R}_{\nu\kappa\lambda\mu} \,. \tag{A.34}$$

Antisymmetry within index pairs

$$\mathcal{R}_{\lambda\mu\nu\kappa} = -\mathcal{R}_{\mu\lambda\nu\kappa} = -\mathcal{R}_{\lambda\mu\kappa\nu} = \mathcal{R}_{\mu\lambda\kappa\nu} \,. \tag{A.35}$$

Index cyclicity

$$\mathcal{R}_{\lambda\mu\nu\kappa} + \mathcal{R}_{\lambda\kappa\mu\nu} + \mathcal{R}_{\lambda\nu\kappa\mu} = 0 \,. \tag{A.36}$$

Related tensors

The Riemann Tensor can be manipulated to produce two unique tensors of lower rank. First, the **Ricci Tensor** (rank-2) can be formed by contraction of the mixed Riemann Tensor (Equation (A.31)) or its fully covariant form (Equation (A.33)):

$$\boxed{\mathcal{R}_{\mu\kappa} = g^{\lambda\nu}\mathcal{R}_{\lambda\mu\nu\kappa} = \mathcal{R}^{\lambda}{}_{\mu\lambda\kappa} \,.} \tag{A.37}$$

By Equation (A.34), this tensor is symmetric in its indices. From the second of these two equalities and Equation (A.31), an explicit form for the Ricci Tensor is

$$\mathcal{R}_{\mu\kappa} = \frac{\partial \Gamma^{\lambda}_{\mu\lambda}}{\partial x^{\kappa}} - \frac{\partial \Gamma^{\lambda}_{\mu\kappa}}{\partial x^{\lambda}} + \Gamma^{\eta}_{\mu\lambda}\Gamma^{\lambda}_{\kappa\eta} - \Gamma^{\eta}_{\mu\kappa}\Gamma^{\lambda}_{\lambda\eta} \,. \tag{A.38}$$

All other contractions of the Riemann Tensor to rank-2 tensors are either equivalent to the Ricci Tensor or are identically 0, so the Ricci Tensor is effectively the only rank-2 tensor derivable from the Riemann Curvature Tensor. The Ricci Tensor can be further contracted to yield the **Scalar Curvature**, a rank-zero (i.e., scalar) tensor:

$$\boxed{\mathcal{R} = g^{\mu\kappa}\mathcal{R}_{\mu\kappa} \,.} \tag{A.39}$$

There does not seem to be a simple expression for the scalar curvature in terms of the metric tensor and affine connection.

Do not let the relative simplicity of the symbols for Ricci Tensor and Scalar Curvature deceive you: written in terms of the metric tensor and affine connection, these are typically very complex mathematical entities (see Equation (7.5)). They are the only tensors of their rank that can be constructed from the metric tensor and its first two derivatives, and that are linear in the second derivatives. They are thus prime candidates for inclusion in the Field Equations of Gravitation.

A.5 Bianchi Identities

The differential properties of the Riemann and associated tensors are important because the Einstein Field Equations, which must incorporate Riemannian curvature, will require divergence-free tensors. In curved space-time the form of derivative that yields tensors is the covariant derivative. Here we do not need to know the details of this creature, which was covered in Section A.3; but need to know only that the covariant derivative with respect to coordinate x^μ is denoted by D_μ and reduces to the ordinary partial derivative in an inertial reference frame, including LIRFs.

In a Locally Inertial Reference Frame (LIRF) – which, according to the Strong Equivalence Principle, can be constructed at any point in space-time – we can find a coordinate system in which the first derivatives of the metric tensor, and hence the affine connection, vanish; so the Riemann Curvature Tensor takes the form (from Equation (A.33)),

$$\mathcal{R}_{\lambda\mu\nu\kappa} = \frac{1}{2}\left[\frac{\partial^2 g_{\lambda\nu}}{\partial x^\mu\,\partial x^\kappa} - \frac{\partial^2 g_{\mu\nu}}{\partial x^\lambda\,\partial x^\kappa} - \frac{\partial^2 g_{\lambda\kappa}}{\partial x^\mu\,\partial x^\nu} + \frac{\partial^2 g_{\mu\kappa}}{\partial x^\lambda\,\partial x^\nu}\right] \qquad \text{(LIRF)}\,,$$

and the covariant derivative is just the normal derivative, $D_\lambda = \partial/\partial x^\lambda$. Then the covariant derivative of the Riemann Tensor in an LIRF is

$$D_\eta \mathcal{R}_{\lambda\mu\nu\kappa} = \frac{1}{2}\left[\overbrace{\frac{\partial^3 g_{\lambda\nu}}{\partial x^\mu\,\partial x^\kappa\partial x^\eta}}^{IV} - \overbrace{\frac{\partial^3 g_{\mu\nu}}{\partial x^\lambda\,\partial x^\kappa\partial x^\eta}}^{I} - \overbrace{\frac{\partial^3 g_{\lambda\kappa}}{\partial x^\mu\,\partial x^\nu\partial x^\eta}}^{II} + \overbrace{\frac{\partial^3 g_{\mu\kappa}}{\partial x^\lambda\,\partial x^\nu\partial x^\eta}}^{III}\right].$$

Now writing the same equation with ν, κ, η varied cyclically:

$$D_\kappa \mathcal{R}_{\lambda\mu\eta\nu} = \frac{1}{2}\left[\overbrace{\frac{\partial^3 g_{\lambda\eta}}{\partial x^\mu\,\partial x^\nu\partial x^\kappa}}^{V} - \overbrace{\frac{\partial^3 g_{\mu\eta}}{\partial x^\lambda\,\partial x^\eta\partial x^\kappa}}^{VI} - \overbrace{\frac{\partial^3 g_{\lambda\nu}}{\partial x^\mu\,\partial x^\kappa\partial x^\eta}}^{IV} + \overbrace{\frac{\partial^3 g_{\mu\nu}}{\partial x^\lambda\,\partial x^\kappa\partial x^\eta}}^{I}\right],$$

$$D_\nu \mathcal{R}_{\lambda\mu\kappa\eta} = \frac{1}{2}\left[\overbrace{\frac{\partial^3 g_{\lambda\kappa}}{\partial x^\mu\,\partial x^\nu\partial x^\eta}}^{II} - \overbrace{\frac{\partial^3 g_{\mu\kappa}}{\partial x^\lambda\,\partial x^\nu\partial x^\eta}}^{III} - \overbrace{\frac{\partial^3 g_{\lambda\eta}}{\partial x^\mu\,\partial x^\nu\partial x^\kappa}}^{V} + \overbrace{\frac{\partial^3 g_{\mu\eta}}{\partial x^\lambda\,\partial x^\eta\partial x^\kappa}}^{VI}\right],$$

and adding the three, the partial derivatives cancel by pairs as indicated and we are left with the *Bianchi Identities:*

$$\boxed{D_\eta \mathcal{R}_{\lambda\mu\nu\kappa} + D_\kappa \mathcal{R}_{\lambda\mu\eta\nu} + D_\nu \mathcal{R}_{\lambda\mu\kappa\eta} = 0\,.} \qquad \text{(A.40)}$$

Since this is manifestly a generally covariant expression it holds in all coordinate systems, not just the LIRF in which we derived it.

A.6 The Einstein Tensor

The covariant divergence of the Riemann Curvature Tensor is not zero, which will be a problem when we come to construct the Einstein Field Equations. We can, however, construct a rank-2, divergence-free tensor from it. We start by contracting the Bianchi Identities with $g^{\lambda\nu}$, recalling that $\mathcal{R}_{....}$ contracts as

$$g^{\lambda\nu}\mathcal{R}_{\lambda\mu\nu\kappa} = \mathcal{R}_{\mu\kappa} \ ,$$

so that

$$g^{\lambda\nu}\mathcal{R}_{\lambda\mu\eta\nu} = -g^{\lambda\nu}\mathcal{R}_{\lambda\mu\nu\eta} = -\mathcal{R}_{\mu\eta}$$

by the antisymmetry condition of Equation (A.35). Since the covariant derivative of the metric tensor is identically zero we can take g inside the derivative operator at will. Thus, applying the contraction $\sum_{\lambda\nu} g^{\lambda\nu}$ to Equation (A.40), we have

$$D_\eta\left(g^{\lambda\nu}\mathcal{R}_{\lambda\mu\nu\kappa}\right) + D_\kappa\left(g^{\lambda\nu}\mathcal{R}_{\lambda\mu\nu\eta}\right) + D_\nu\left(g^{\lambda\nu}\mathcal{R}_{\lambda\mu\kappa\eta}\right) = 0 \ .$$

The first two contractions are given by the previous two equations, the last by $D_\eta\mathcal{R}^{\nu}{}_{\mu\kappa\eta}$, so

$$D_\eta\mathcal{R}_{\mu\kappa} - D_\kappa\mathcal{R}_{\mu\eta} + D_\nu\mathcal{R}^{\nu}{}_{\mu\kappa\eta} = 0 \ .$$

Now contract again with $g^{\mu\kappa}$, once again taking the metric tensor within the derivative:

$$D_\eta\left(g^{\mu\kappa}\mathcal{R}_{\mu\kappa}\right) - D_\kappa\left(g^{\mu\kappa}\mathcal{R}_{\mu\eta}\right) - D_\nu\left(g^{\mu\kappa}\mathcal{R}^{\nu}{}_{\mu\eta\kappa}\right) = 0$$

(where we have again invoked Equation (A.35) in the last term), or

$$D_\eta\mathcal{R} - D_\kappa\mathcal{R}^{\kappa}_{\eta} - D_\nu\mathcal{R}^{\nu}_{\eta} = 0 \ .$$

Combining dummy indices,

$$D_\eta\mathcal{R} - 2D_\kappa\mathcal{R}^{\kappa}_{\eta} = 0 \ .$$

Now, $D_\tau\mathcal{R} = D_\lambda\left(\delta^{\lambda}_{\tau}\mathcal{R}\right)$ so we can write the above expression with a common derivative:

$$D_\mu\left(\delta^{\mu}_{\eta}\mathcal{R} - 2\mathcal{R}^{\mu}_{\eta}\right) = 0 \ .$$

Now we lower indices with $g_{\mu\nu}$ (again taking it inside the derivative):

$$\begin{aligned} g_{\mu\nu} D_\mu \left(\delta^\mu_\eta \mathcal{R} - 2\mathcal{R}^\mu_\eta \right) &= D_\mu \left(g_{\mu\nu} \delta^\mu_\eta \mathcal{R} - 2 g_{\mu\nu} \mathcal{R}^\mu_\eta \right) , \\ &= D_\mu \left(g_{\nu\eta} \mathcal{R} - 2\mathcal{R}_{\nu\eta} \right) , \\ &= 0 . \end{aligned}$$

It follows that the quantity $g_{\nu\eta}\mathcal{R} - 2\mathcal{R}_{\nu\eta}$ has zero derivatives with respect to all coordinates, and is manifestly a tensor (since all its components are) and so is divergence-free in all coordinate systems. This combination turns out to be the simplest non-trivial, divergence-free, rank-2 tensor that is derivable from the Riemann Curvature Tensor and so was chosen by Einstein for the curvature component of his Field Equations. It is commonly known as the *Einstein Tensor*, denoted (here) by $\mathcal{G}$ and usually written as

$$\boxed{\mathcal{G}_{\mu\nu} \equiv \mathcal{R}_{\mu\nu} - \tfrac{1}{2} g_{\mu\nu} \mathcal{R} .} \qquad \text{(A.41)}$$

Appendix B

Newtonian approximations

Equations of Motion

The quasi-Newtonian limitations on the gravitational field introduced in Chapter 5:

1. all velocities are small compared to that of light,
2. the gravitational field is stationary (not changing with time),
3. the gravitational field is very weak,

allow us to reduce the GR Equations of Motion (5.9) to the Newtonian form (5.11) in six steps. We begin by postulating that **g** is diagonal (recall: it is always possible to find a coordinate system in which **g** is diagonal). Then:

1. **Small velocities** $\left|dx^1, dx^2, dx^3\right| \ll |cdt| = \left|dx^0\right| \Rightarrow dx^i/d\tau \ll dx^0/d\tau$ for all $i = 1, 2, 3$, so we need only consider $\mu = \nu = 0$ in the Equations of Motion, Equation (5.9):

$$\frac{d^2x^\lambda}{d\tau^2} = -\Gamma^\lambda_{00}\left(\frac{dx^0}{d\tau}\right)^2 \quad \text{for all } \lambda = 0, 1, 2, 3\ , \tag{B.1}$$

   where $x^0 = ct$.

2. **Stationary field** $\partial g_{\alpha\beta}/\partial\,(ct) = \partial g_{\alpha\beta}/\partial x^0 = 0$ for all $\alpha, \beta \Rightarrow$ the non-zero components of the affine connection become

$$\begin{aligned}\Gamma^\lambda_{00} &= \frac{1}{2}g^{\kappa\lambda}\left[\overbrace{\frac{\partial g_{0\kappa}}{\partial x^0}}^{0} + \overbrace{\frac{\partial g_{0\kappa}}{\partial x^0}}^{0} - \frac{\partial g_{00}}{\partial x^\kappa}\right], \\ &= -\frac{1}{2}g^{\kappa\lambda}\frac{\partial g_{00}}{\partial x^\kappa}\ .\end{aligned}$$

Since g is diagonal the implied sum over κ collapses to

$$\Gamma^{\lambda}_{00} = -\frac{1}{2}g^{\lambda\lambda}\frac{\partial g_{00}}{\partial x^{\lambda}} \qquad \text{(no sum on } \lambda \text{ !)}$$

for all $\lambda = 0, 1, 2, 3$. Combining this with Equation (B.1), the simplified Equations of Motion are

$$\frac{d^2x^{\lambda}}{d\tau^2} = \frac{1}{2}\left(\frac{dx^0}{d\tau}\right)^2 g^{\lambda\lambda}\frac{\partial g_{00}}{\partial x^{\lambda}} \quad \text{for all } \lambda \quad \text{(no sum on } \lambda) . \qquad \text{(B.2)}$$

3. **Weak field** We assume that the metric tensor differs from the LIRF (Minkowski) metric tensor by only small amounts: $g_{\lambda\lambda} = \eta_{\lambda\lambda} + h_{\lambda\lambda} = (\pm 1 + h_{\lambda\lambda})$, where $|h_{\lambda\lambda}| \ll 1$ for all λ and we use -1 only for $\lambda = 0$ (since $\eta = \text{diag}(-1, 1, 1, 1)$). From this we deduce

$$1\colon g_{00} = -1 + h_{00} \; ;$$

$$2\colon \frac{\partial g_{00}}{\partial x^{\lambda}} = \frac{\partial h_{00}}{\partial x^{\lambda}} \quad \text{for all } \lambda \; .$$

In addition, we note that $g^{\lambda\lambda} = 1/g_{\lambda\lambda} = 1/\left(\pm 1 + h_{\lambda\lambda}\right) \approx \pm 1 - h_{\lambda\lambda}$ since $\mathbf{g}$ is diagonal and $|h_{\lambda\lambda}| \ll 1$. Then

$$g^{\lambda\lambda}\frac{\partial g_{00}}{\partial x^{\lambda}} \approx (\pm 1 - h_{\lambda\lambda})\frac{\partial h_{00}}{\partial x^{\lambda}} \quad \text{for all } \lambda \quad \text{(no sum on } \lambda)$$

(where -1 is used only for $\lambda = 0$). Substituting this result into Equation (B.2) the Equations of Motion become

$$\frac{d^2x^{\lambda}}{d\tau^2} = \frac{1}{2}\left(\frac{dx^0}{d\tau}\right)^2 (\pm 1 - h_{\lambda\lambda})\frac{\partial h_{00}}{\partial x^{\lambda}} \quad \text{for all } \lambda \quad \text{(no sum on } \lambda) . \quad \text{(B.3)}$$

Equation (B.3) is the form of particle trajectories under the three classical assumptions made at the start of this section. It has the following consequences.

4. **Time trajectory** Putting $\lambda = 0$ and $dx^0 = cdt$ in the above expression:

$$\begin{aligned} c^2\frac{d^2t}{d\tau^2} &= \frac{c^2}{2}\left(\frac{dt}{d\tau}\right)^2 (-1 - h_{00})\frac{\partial h_{00}}{\partial x^0} \\ &= 0 \, , \end{aligned}$$

since the field is stationary (so $\partial h_{00}/\partial x^0 = 0$). Then

$$\frac{d^2t}{d\tau^2} = 0 \Rightarrow \frac{dt}{d\tau} = \text{constant}, \qquad \text{(B.4)}$$

so ct scales directly with τ in the Newtonian approximation.

5. **Space trajectory** For $\lambda = i = 1, 2, 3$, Equation (B.3) becomes

$$\frac{d^2x^i}{d\tau^2} = \frac{c^2}{2}\left(\frac{dt}{d\tau}\right)^2 (1 - h_{ii})\frac{\partial h_{00}}{\partial x^i} \approx \frac{c^2}{2}\left(\frac{dt}{d\tau}\right)^2 \frac{\partial h_{00}}{\partial x^i}$$

to first order in h, since $h_{ii}\left(\partial h_{00}/\partial x^i\right)$ is much smaller than $\partial h_{00}/\partial x^i$ and so can be ignored. Since $dt/d\tau$ is a constant by Equation (B.4), we can divide both sides of the above equation by $(dt/d\tau)^2$ to obtain true space trajectories:

$$\frac{d^2x^i}{dt^2} = \frac{c^2}{2}\frac{\partial h_{00}}{\partial x^i} \quad \text{for all } i = 1, 2, 3\ .$$

Writing this in 'normal' 3-vector notation,

$$\frac{d^2\vec{x}}{dt^2} = \frac{c^2}{2}\vec{\nabla} h_{00}\ . \tag{B.5}$$

6. **Comparison with Newtonian gravity** Comparing Equation (B.5) with the Newtonian form, $d^2\vec{x}/dt^2 = -\vec{\nabla}\Phi$, gives $\vec{\nabla} h_{00} = -2\vec{\nabla}\Phi/c^2 \Rightarrow h_{00} = -2\Phi/c^2 +$ constant. Since both h_{00} and Φ go to 0 at infinity, the constant is zero and thus $h_{00} = -2\Phi/c^2$, so the metric tensor satisfies

$$g_{00} \approx -\left(1 + \frac{2\Phi}{c^2}\right) \tag{B.6}$$

in the Newtonian approximation.

Gravitational Field Equations

We show here that the Einstein Field Equations reduce to those of Newtonian gravitation in the static, weak-field limit. We start with the general form of the Field Equations (7.10):

$$\mathcal{R}_{\mu\nu} = \kappa\left[T_{\mu\nu} - \frac{1}{2}g_{\mu\nu}T\right] ,$$

and note that the rest energy will be much greater than any other energy in this limit, so we need consider only the leading term

$$\mathcal{R}_{00} = \kappa\left[T_{00} - \frac{1}{2}g_{00}T\right] .$$

Now,

$$T = g^{\mu\nu}T_{\mu\nu} \approx g^{00}T_{00} = \frac{1}{g_{00}}g_{00} = 1$$

since g will be diagonal (or can be diagonalized) in Newtonian space-time, and $T_{00} \approx g_{00}$. Then the Field Equations reduce to

$$\mathcal{R}_{00} \approx \frac{1}{2}\kappa\rho c^2 . \tag{B.7}$$

To go further we need to develop an alternative expression for $\mathcal{R}_{00}$ arising from curvature. From the definition of the Ricci Tensor (Equation (A.37)),

$$\mathcal{R}_{00} = g^{\mu\nu}\mathcal{R}_{\mu 0\nu 0} .$$

From symmetry properties of the Riemann Tensor (Equations (A.34) and (A.36)), we have $\mathcal{R}_{\mu 000} = \mathcal{R}_{00\mu 0} = 0$ for all indices μ, so the above equation reduces to

$$\mathcal{R}_{00} = g^{ij}\mathcal{R}_{i0j0} , \tag{B.8}$$

for $i,j = 1,2,3$.

Now, in the weak-field limit we can approximate the Riemann Curvature Tensor (Equation (A.33)) with only the lowest powers of the metric tensor components:

$$\mathcal{R}_{\mu\nu\kappa\lambda} \approx \frac{1}{2}\left(\frac{\partial^2 g_{\nu\lambda}}{\partial x^\mu \partial x^\kappa} - \frac{\partial^2 g_{\mu\lambda}}{\partial x^\nu \partial x^\kappa} + \frac{\partial^2 g_{\mu\kappa}}{\partial x^\nu \partial x^\lambda} - \frac{\partial^2 g_{\nu\kappa}}{\partial x^\mu \partial x^\lambda}\right) .$$

Substituting this into Equation (B.8),

$$\mathcal{R}_{00} \approx \frac{g^{ij}}{2}\left(\frac{\partial^2 g_{00}}{\partial x^i \partial x^j} - \frac{\partial^2 g_{i0}}{\partial x^0 \partial x^j} + \frac{\partial^2 g_{0j}}{\partial x^i \partial x^0} - \frac{\partial^2 g_{ij}}{\partial x^0 \partial x^0}\right) .$$

For static fields $\partial/\partial x^0 = 0$, so this reduces to

$$\mathcal{R}_{00} \approx \frac{g^{ij}}{2}\frac{\partial^2 g_{00}}{\partial x^i \partial x^j} ,$$

with $i,j \neq 0$. Again invoking weak fields, $g^{ij} = 1 + h^{ij}$ with $|h^{ij}| \ll 1$, and since g is diagonal we have, to first order, $g^{ij} \approx \delta^{ij}$ which is 1 if $i = j$ and zero otherwise:

$$\mathcal{R}_{00} \approx \frac{\delta^{ij}}{2}\frac{\partial^2 g_{00}}{\partial x^i \partial x^j} = \frac{1}{2}\nabla^2 g_{00} . \tag{B.9}$$

Now, we have already shown that the trajectory of freely falling particles reduces to the Newtonian result in the static, weak-field limit only if (Equation (B.6))

$$g_{00} \approx -\left(1 + 2\frac{\Phi}{c^2}\right) ,$$

so that

$$\mathcal{R}_{00} \approx -\frac{1}{c^2}\nabla^2\Phi \, .$$

Comparing this with Equation (B.7),

$$\nabla^2\Phi \approx -\frac{1}{2}\kappa\rho c^4 \, .$$

The Newtonian Field Equation is $\nabla^2\Phi = 4\pi G\rho$, so that we must have

$$\kappa = -\frac{8\pi G}{c^4} \tag{B.10}$$

if the Einstein Field Equations are to be compatible with Newtonian gravity in the static, weak-field limit. This justifies the choice of κ made in the Field Equations, Equation (7.4).

Thus, with this choice of normalizing constant the Einstein Field Equations reduce to those of Newtonian gravity in the appropriate limit, and therefore are reasonable choices for the field equations without cosmological constant. This will also be true with a non-zero value for Λ, providing that it is very small.

Appendix C

Useful numbers

Common cosmological units

Distance:	Gigaparsec (Gpc)	=	3.086×10^{25} m
Time:	Gigayear (Gyr)	=	3.156×10^{16} sec
Energy:	Mega-electron-Volt (MeV)	=	1.602×10^{-13} J

Physical constants

Speed of light:	c	=	2.9979×10^{8} m/sec
		=	0.3067 Gpc/Gyr (or Mpc/Myr, etc.)
Gravitational constant:	G	=	6.673×10^{-11} m^3/kg-sec^2
Planck's constant:	h	=	6.626×10^{-34} J-sec
	$\hbar$	=	1.055×10^{-34} J-sec
Boltzmann's constant:	k_{B}	=	1.381×10^{-23} J/K
		=	8.617×10^{-5} eV/K
Baryon mass:[1]	m_{B}	$\approx$	1.674×10^{-27} kg
		$\approx$	939 MeV/c^2
Stefan–Boltzmann constant:	σ_{SB}	=	5.671×10^{-8} W/m^2/K^4
Radiation density constant:	a_{rad}	=	7.566×10^{-16} J/m^3/K^4
Photon density constant:	β	=	2.030×10^{7} photons/m^3/K^3

[1] Mean of proton and neutron masses

Astronomical constants

Solar mass:	$M_\odot$	=	1.989×10^{30} kg
Solar luminosity:	$L_\odot$	=	3.846×10^{26} W
Astronomical Unit:	AU	=	1.496×10^{11} m
Parsec:	pc	=	3.086×10^{16} m
		=	206,265 AU
		=	3.262 light years
Year (tropical):	yr	=	3.156×10^{7} sec

Cosmological parameters

Estimated current values

Hubble Constant:	H_0	$\approx$	$0.074\ \text{Gyr}^{-1}$
		$\approx$	$2.3 \times 10^{-18}\ \text{sec}^{-1}$
		$\approx$	72 km/sec/Mpc
Critical mass/energy density:	$\rho_{c,0}$	$\approx$	$9.8 \times 10^{-27}\ \text{kg/m}^3$
		$\approx$	$1.5 \times 10^{11}\ \text{M}_\odot/\text{Mpc}^3$
	$\varepsilon_{c,0}$	$\approx$	$8.8 \times 10^{-10}\ \text{J/m}^3$
		$\approx$	$5500\ \text{MeV/m}^3$
CMB temperature:	$T_{\text{CMB},0}$	$\approx$	2.725 K
CMB energy density:	$\varepsilon_{\text{CMB},0}$	$\approx$	$4.17 \times 10^{-14}\ \text{J/m}^3$
Primordial neutrino density:	$\varepsilon_{\nu,0}$	$\approx$	$2.84 \times 10^{-14}\ \text{J/m}^3$
Visible light energy density:	$\varepsilon_{\lambda,0}$	$\approx$	$3.5 \times 10^{-16}\ \text{J/m}^3$
Baryon/photon number ratio:	η	$\approx$	5.5×10^{-10}

Planck units

Planck time:	t_P	=	$\left(G\hbar/c^5\right)^{1/2}$	$= 5.39 \times 10^{-44}$ s
Planck mass:	M_P	=	$(\hbar c/G)^{1/2}$	$= 2.18 \times 10^{-8}$ kg
Planck length:	l_P	=	$\left(G\hbar/c^3\right)^{1/2}$	$= 1.62 \times 10^{-35}$ m
Planck energy:	E_P	=	$\left(\hbar c^5/G\right)^{1/2}$	$= 1.96 \times 10^{9}$ J
				$= 1.22 \times 10^{19}$ GeV
Planck temperature:	T_P	=	$\left(\hbar c^5 / G k_B^2\right)^{1/2}$	$= 1.42 \times 10^{32}$ K

Appendix D

Symbols

Times

t_0	Current time	Equation (9.32)
t_e	Photon emission	Equation (9.33)
t_{gc}	Gravitational contraction time	Section 10.4.2
t_H	Hubble Time	Equation (9.2)
t_{ie}	Inflation ends	$\sim 10^{-34}$ s (nominal)
t_{is}	Inflation starts	$\sim 10^{-36}$ s (nominal)
t_{lb}	Look-back time	Equation (9.34)
t_{pd}	Photon decoupling ('recombination')	$\approx$ 375 kyr*
t_{nuc}	Primordial nucleosynthesis begins	$\approx$ 100 s
t_r	Photon reception	
t_{rm}	Radiation-matter equality	$\approx$ 52 kyr

*Listed times are currently estimated CCM values

Distances

d_A	=	Angular diameter distance	Equation (9.72)
d_{cm}	=	Co-moving distance	Equation (9.54)
d_{EH}	=	Event horizon proper distance	Equation (10.12)
d_H	=	Hubble distance	Equation (9.3)
d_L	=	Luminosity distance	Equation (9.61)
d_p	=	Proper distance	Section 9.2.3
d_0	=	Current proper distance	Equation (9.44)
d_{PH}	=	Particle horizon proper distance	Equation (10.9)
d_{SH}	=	Sound horizon proper distance	Equation (10.14)

Tensors and related quantities

D	Covariant derivative	Appendix A.3
$\mathcal{G}$	Einstein Tensor	Section 7.2.1, Appendix A.6

$\mathbf{g}$, $g_{..}$	Metric tensor	Sections 2.3.1, 4.2
$\Gamma^{\cdot}_{..}$	Affine connection	Equations (5.6), (5.8), Appendix A.1
$\mathcal{R}^{\cdot}_{...}$	Riemann Curvature Tensor	Equation (6.10), Appendix A.4
$\mathcal{R}_{..}$	Ricci Tensor	Equation (6.12), Appendix A.4
$\mathcal{R}$	Curvature scalar	Equation (6.13), Appendix A.4
$\mathcal{T}_{..}$	Energy–momentum tensor	Sections 7.1.1, 8.3.1

Others

a	Expansion parameter	Sections 3.1, 8.1.2
a	Acceleration	Chapter 1, Section 12.3
ε, $\varepsilon_.$	Energy density	Section 8.3
ε_c	Critical energy density	Equation (9.7)
H, H_0	Hubble Parameter	Equations (9.1), (9.4)
h	Hubble parameter scale	Section 9.1.1
	Disk scale height	Section 12.3.1
μ_0	Distance modulus	Section 9.3.1
η	Baryon/photon number ratio	Equation (11.3), Sections 11.3, 16.3.2
	Conformal time	Equation (9.39)
	Minkowski metric	Section 8.2.1
K, K_0	Geometric curvature	Section 9.1.3
Λ	Cosmological Constant	Sections 8.3.1, 13.2
q, q_0	Deceleration parameter	Section 9.3.4
R, R_0	Radius of curvature	Section 6.1.2, Equation (9.49)
r, r_g	Co-moving radial coordinate	Section 8.1.2, Equation (9.56)
w	Equation-of-state parameter	Equations (8.25), (8.26)
Ω, $\Omega_.$	Energy density parameter	Section 9.1.2
z	Redshift	Section 9.2.1, Equation (9.28)

Abbreviations and acronyms

BAO	Baryonic acoustic oscillation	Section 18.4
CCM	Concordance Cosmological Model	Chapter 15
CMB	Cosmological Microwave Background Radiation	Section 11.1
EOS	Equation of State	Section 8.3.3
GR	General Relativity	
LTB	Lemaître–Tolman–Bondi metric	Section 13.3.2
LIRF	Locally Inertial Reference Frame	Section 5.2
RW	Robertson–Walker metric	Section 8.1.4
SEP	Strong Equivalence Principle	Section 5.2
SNF	Strong nuclear force	
SR	Special Relativity	
WEP	Weak Equivalence Principle	Section 5.1
WNF	Weak nuclear force	

References

Amendola, L. and Tsujikawa, S. 2010, *Dark Energy*, Cambridge University Press

Bernstein, J. 1995, *An Introduction to Cosmology*, Prentice-Hall

Bertin, G. 2000, *Dynamics of Galaxies*, Cambridge University Press

Carroll, S. M. 2004, *Spacetime and Geometry*, Addison-Wesley

Carroll, B. W. and Ostlie, D. A. 2007, *An Introduction to Modern Astrophysics* (2nd Edn.), Pearson/Addison-Wesley

Cheng, T.-P. 2005, *Relativity, Gravitation and Cosmology*, Oxford University Press

Collier, P. 2013, *A Most Incomprehensible Thing*, Incomprehensible Books

Ellis, G. F. R., Maarten, R., and MacCallum, M. A. H. 2012, *Relativistic Cosmology*, Cambridge University Press

Ferreira, P. D. 2014, *The Perfect Theory*, Houghton Mifflin Harcourt

Ghosh, A. 2000, *Origins of Inertia*, Apeiron

Graves, J. C. 1971, *The Conceptual Foundations of Contemporary Relativity Theory*, MIT Press

Guth, A. H. 1981, *Phys. Rev. D* **34**, 347

Hartle, J. B. 2003, *Gravitation*, Addison-Wesley

Hobson, M. P., Efstathiou, G. P. and Lasenby, A. N. 2006, *General Relativity: an Introduction for Physicists*, Cambridge University Press

Kolb, E. W. and Turner, M. S. 1990, *The Early Universe*, Addison-Wesley

Liddle, A. R. and Lyth, D. H. 2000, *Cosmological Inflation and Large-Scale Structure*, Cambridge University Press

Loeb, A. and Furlanetto, S. R. 2013, *The First Galaxies in the Universe*, Princeton University Press

Longair, M. S. 2008, *Galaxy Formation* (2nd Edn.), Springer

Mo, H., van den Bosch, F., and White, S. 2010, *Galaxy Formation and Evolution*, Cambridge University Press

Narlikar, J. V. 1993, *Introduction to Cosmology* (2nd Edn.), Cambridge University Press

Nussbaumer, H. and Bieri, L. 2009, *Discovering the Expanding Universe*, Cambridge University Press

Ostriker, J. P. and Mitton, S. 2013, *Heart of Darkness*, Princeton University Press

Peacock, J. A. 1999, *Cosmological Physics*, Cambridge University Press

Peebles, P. J. E. 1993, *Principles of Physical Cosmology*, Princeton University Press

Peter, P. and Uzan, J.-P. 2009, *Primordial Cosmology*, Oxford University Press

Rees, M. 1995, *Perspectives in Astrophysical Cosmology*, Cambridge University Press

Rindler, W. 1977, *Essential Relativity* (2nd Edn., Revised), Springer-Verlag

Ryden, B. 2003, *Introduction to Cosmology*, Addison-Wesley

Sanders, R. H. 2010, *The Dark Matter Problem*, Cambridge University Press

Schneider, P. 2010, *Extragalactic Astronomy and Cosmology*, Springer

Sciama, D. W. 1969, *The Physical Foundations of General Relativity*, Doubleday

Serjeant, S. 2010, *Observational Cosmology,* Cambridge University Press

Sparke, L. S. and Gallagher, J. S. III 2007, *Galaxies in the Universe* (2nd Edn.), Cambridge University Press

Unger, R. M. and Smolin, L. 2015, *The Singular Universe and the Reality of Time*, Cambridge University Press

Weinberg, S. 2008, *Cosmology,* Oxford University Press

Index